Discovering Partial Least Squares with JMP®

Ian Cox and Marie Gaudard

support.sas.com/bookstore

The correct bibliographic citation for this manual is as follows: Cox, Ian and Gaudard, Marie. 2013. *Discovering Partial Least Squares with JMP®*. Cary, NC: SAS Institute Inc.

Discovering Partial Least Squares with JMP®

Copyright © 2013, SAS Institute Inc., Cary, NC, USA

ISBN 978-1-61290-829-8 (electronic book)
ISBN 978-1-61290-822-9

All rights reserved. Produced in the United States of America.

For a hard-copy book: No part of this publication may be reproduced, stored in a retrieval system, or transmitted, in any form or by any means, electronic, mechanical, photocopying, or otherwise, without the prior written permission of the publisher, SAS Institute Inc.

For a web download or e-book: Your use of this publication shall be governed by the terms established by the vendor at the time you acquire this publication.

The scanning, uploading, and distribution of this book via the Internet or any other means without the permission of the publisher is illegal and punishable by law. Please purchase only authorized electronic editions and do not participate in or encourage electronic piracy of copyrighted materials. Your support of others' rights is appreciated.

U.S. Government License Rights; Restricted Rights: The Software and its documentation is commercial computer software developed at private expense and is provided with RESTRICTED RIGHTS to the United States Government. Use, duplication or disclosure of the Software by the United States Government is subject to the license terms of this Agreement pursuant to, as applicable, FAR 12.212, DFAR 227.7202-1(a), DFAR 227.7202-3(a) and DFAR 227.7202-4 and, to the extent required under U.S. federal law, the minimum restricted rights as set out in FAR 52.227-19 (DEC 2007). If FAR 52.227-19 is applicable, this provision serves as notice under clause (c) thereof and no other notice is required to be affixed to the Software or documentation. The Government's rights in Software and documentation shall be only those set forth in this Agreement.

SAS Institute Inc., SAS Campus Drive, Cary, North Carolina 27513-2414.

October 2013

SAS provides a complete selection of books and electronic products to help customers use SAS® software to its fullest potential. For more information about our offerings, visit **support.sas.com/bookstore** or call 1-800-727-3228.

SAS® and all other SAS Institute Inc. product or service names are registered trademarks or trademarks of SAS Institute Inc. in the USA and other countries. ® indicates USA registration.

Other brand and product names are trademarks of their respective companies.

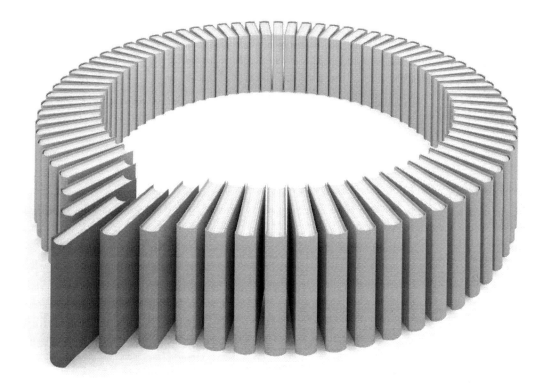

Gain Greater Insight into Your JMP® Software with SAS Books.

Discover all that you need on your journey to knowledge and empowerment.

support.sas.com/bookstore
for additional books and resources.

SAS and all other SAS Institute Inc. product or service names are registered trademarks or trademarks of SAS Institute Inc. in the USA and other countries. ® indicates USA registration. Other brand and product names are trademarks of their respective companies. © 2013 SAS Institute Inc. All rights reserved. S108082US.0613

Contents

Preface xi
A Word to the Practitioner.. xi
The Organization of the Book xi
Required Software xii
Accessing the Supplementary Content xii

Chapter 1 Introducing Partial Least Squares 1
Modeling in General 1
Partial Least Squares in Today's World 2
Transforming, and Centering and Scaling Data 3
An Example of a PLS Analysis 4
    The Data and the Goal... 4
    The Analysis 5
    Testing the Model 9

Chapter 2 A Review of Multiple Linear Regression 11
The Cars Example 11
Estimating the Coefficients... 15
Underfitting and Overfitting: A Simulation 16
The Effect of Correlation among Predictors: A Simulation 18

Chapter 3 Principal Components Analysis: A Brief Visit 25
Principal Components Analysis 25
Centering and Scaling: An Example 25

The Importance of Exploratory Data Analysis in Multivariate Studies 31
Dimensionality Reduction via PCA 34

Chapter 4 A Deeper Understanding of PLS 37
Centering and Scaling in PLS 37
PLS as a Multivariate Technique 38
Why Use PLS? 39
How Does PLS Work? 45
PLS versus PCA 49
PLS Scores and Loadings 50
    Some Technical Background 50
An Example Exploring Prediction 59
    One-Factor NIPALS Model 60
    Two-Factor NIPALS Model 63
    Variable Selection 64
    SIMPLS Fits 64
Choosing the Number of Factors 65
    Cross Validation 65
    Types of Cross Validation 66
    A Simulation of K-Fold Cross Validation 69
    Validation in the PLS Platform 69
The NIPALS and SIMPLS Algorithms 71
Useful Things to Remember About PLS 72

Chapter 5 Predicting Biological Activity 75
Background 75
The Data 76
    Data Table Description 76
    Initial Data Visualization 77
A First PLS Model 79
    Our Plan 79
    Performing the Analysis 79
    The Partial Least Squares Report 81
    The SIMPLS Fit Report 82
    Other Options 83
A Pruned PLS Model 93

Model Fit	93
Diagnostics	95
Performance on Data from Second Study	96
Comparing Predicted Values for the Second Study to Actual Values	96
Comparing Residuals for Both Studies	99
Obtaining Additional Insight	101
Conclusion	104

Chapter 6 Predicting the Octane Rating of Gasoline 105

Background 105
The Data 106
    Data Table Description 106
    Creating a Test Set Indicator Column 107
Viewing the Data 108
    Octane and the Test Set 108
    Creating a Stacked Data Table 109
    Constructing Plots of the Individual Spectra 111
    Individual Spectra 112
    Combined Spectra 113
A First PLS Model 116
    Excluding the Test Set 116
    Fitting the Model 117
    The Initial Report 118
A Second PLS Model 120
    Fitting the Model 120
    High-Level Overview 120
    Diagnostics 121
    Score Scatterplot Matrices 125
    Loading Plots 127
    VIPs 129
    Model Assessment Using Test Set 133
A Pruned Model 136

Chapter 7 Equation Chapter 1 Section 1Water Quality in the Savannah River Basin 139

Background 140
The Data 141

Data Table Description	141
Initial Data Visualization	144
Missing Response Values	145
Impute Missing Data	146
Distributions	147
Transforming AGPT	148
Differences by Ecoregion	150
Conclusions from Visual Analysis and Implications	155
A First PLS Model for the Savannah River Basin	155
Our Plan	155
Performing the Analysis	156
The Partial Least Squares Report	159
The NIPALS Fit Report	159
Defining a Pruned Model	163
A Pruned PLS Model for the Savannah River Basin	166
Model Fit	166
Diagnostics	168
Saving the Prediction Formulas	169
Comparing Actual Values to Predicted Values for the Test Set	170
A First PLS Model for the Blue Ridge Ecoregion	173
Making the Subset	173
Reviewing the Data	174
Performing the Analysis	175
The NIPALS Fit Report	176
A Pruned PLS Model for the Blue Ridge Ecoregion	178
Model Fit	178
Comparing Actual Values to Predicted Values for the Test Set	179
Conclusion	181
Chapter 8 Baking Bread That People Like	**183**
Background	183
The Data	184
Data Table Description	184
Missing Data Check	186
The First Stage Model	187
Visual Exploration of Overall Liking and Consumer Xs	187

  The Plan for the First Stage Model 189
  Stage One PLS Model 190
  Stage One Pruned PLS Model 195
  Stage One MLR Model 197
  Comparing the Stage One Models 200
  Visual Exploration of Ys and Xs 202
  Stage Two PLS Model 207
  Stage Two MLR Model 212
The Combined Model for Overall Liking 215
  Constructing the Prediction Formula 215
  Viewing the Profiler 218
Conclusion 219

Appendix 1: Technical Details 221

Ground Rules 222
The Singular Value Decomposition of a Matrix 222
  Definition 222
  Relationship to Spectral Decomposition 223
  Other Useful Facts 223
Principal Components Regression 223
The Idea behind PLS Algorithms 224
NIPALS 225
  The NIPALS Algorithm 225
  Computational Results 228
  Properties of the NIPALS Algorithm 231
SIMPLS 237
  Optimization Criterion 237
  Implications for the Algorithm 237
  The SIMPLS Algorithm 238
More on VIPs 244
The Standardize X Option 246
Determining the Number of Factors 246
  Cross Validation: How JMP Does It 246

Appendix 2: Simulation Studies 249

Introduction 249
The Bias-Variance Tradeoff in PLS 250

Introduction	250
Two Simple Examples	250
Motivation	254
The Simulation Study	255
Results and Discussion	257
Conclusion	261
Using PLS for Variable Selection	**263**
Introduction	263
Structure of the Study	264
The Simulation	267
Computation of Result Measures	268
Results	270
Conclusion	280
References	**281**
Index	**285**

Preface

A Word to the Practitioner

Welcome to *Discovering Partial Least Squares with JMP*. This book introduces you to the exciting area of partial least squares. Partial least squares is a multivariate modeling technique based on the idea of projection—the inspiration for the book's cover design. You will obtain background understanding and see the technique applied in a number of examples. The book is built around the intuitive and powerful JMP statistical software, which will help you understand and internalize this new topic in a way that just reading simply cannot.

Since our goal is to help you apply partial least squares in your own setting, the textual material exists only to build your understanding and confidence as you progress through the worked examples. Although we endeavor to provide the salient details, the area of partial least squares is very broad and this book is necessarily incomplete. To the extent that we cannot cover certain topics fully, we provide references for your further study.

The Organization of the Book

We open with a number of introductory chapters that describe the concepts behind partial least squares and help position it in the wider world of statistical methodology and application. The meat of the book is found in Chapters 5 through 8, which contain four examples. Working through these examples using JMP prepares you to apply partial least squares to your own data. The book also contains two appendixes that provide further statistical details and the results of some simulation studies. Depending on your level and area of interest, you might find these useful.

xii *Preface*

Required Software

Although a user of standard JMP 11 or later will find this book useful, many examples require JMP Pro 11 or later. Compared to the standard version of JMP, the Pro version is intended for those who require deeper analytical capabilities. In JMP Pro, the implementation of partial least squares is quite complete.

The book uses JMP Pro 11.0 in screenshots, instructions, and discussions. Even though JMP's PLS capabilities will continue to be developed, the major features and design shown here will persist. However, in future versions, you may notice very slight differences from the specific instruction sequences and screenshots presented in this book.

Ideally, you will have JMP Pro 11 available as you work through this book. A fully functional version of JMP Pro 11 that runs for 30 days can be requested at http://www.jmp.com/webforms/jmp_pro_eval.shtml.

The standard version of JMP enables you to run some partial least squares analyses through a simplified interface. Using this version you will be able to work through some, but not all, of the examples, and many of the scripts linked to in the book will not function correctly. But the book should still help your understanding of partial least squares, and help you decide if you need the Pro version of JMP.

Accessing the Supplementary Content

The data tables and scripts associated with the book can be accessed at either http://support.sas.com/cox or http://support.sas.com/gaudard, which provides a single ZIP file. Once downloaded, you can unzip the contents to a convenient location on your hard disk. This process creates a master JMP journal file Discovering Partial Least Squares with JMP.jrn, along with a folder for each chapter containing scripts. Data tables are created by running these scripts using the links in the master journal. The master journal file provides a convenient way to access all of the supplementary content, and the instructions in the text assume that you will do this.

The data tables themselves contain saved scripts that are referred to in the chapters. Often, when working through an example, we show the steps that you can follow to generate a report in JMP. In addition, either parenthetically or directly, we give the name of a script that has been saved to the data table and that generates that same analysis.

This way, if you want to see the report without stepping through the selections to create it, you can simply run that script.

The scripts are used to illustrate concepts and to help you develop understanding. Because many of the scripts have an element of randomness built in, it is usually worth running the same script more than once to see the effect over various random choices. Also, be aware that the scripts have been encrypted. If you open one of these scripts directly rather than via the journal file mentioned earlier, you see what appears to be gibberish. Nevertheless, you can right-click within the script window and select **Run Script**.

Introducing Partial Least Squares

Modeling in General 1
Partial Least Squares in Today's World 2
Transforming, and Centering and Scaling Data 3
An Example of a PLS Analysis 4
    The Data and the Goal 4
    The Analysis 5
    Testing the Model 9

Modeling in General

Applied statistics can be thought of as a body of knowledge, or even a technology, that supports learning about the real world in the face of uncertainty. The theme of *learning* is ubiquitous in more or less every context that can be imagined, and along with this comes the idea of a (statistical) *model* that tries to codify or encapsulate our current understanding.

Many statistical models can be thought of as relating one or more *inputs* (which we call collectively **X**) to one or more *outputs* (collectively **Y**). These quantities are measured on the items or units of interest, and models are constructed from these observations. Such observations yield quantitative data that can be expressed numerically or coded in numerical form.

By the standards of fundamental physics, chemistry, and biology, at least, statistical models are generally useful when current knowledge is moderately low and the underlying mechanisms that link the values in **X** and **Y** are obscure. So although one of the perennial challenges of any modeling activity is to take proper account of whatever is already known, the fact remains that statistical models are generally *empirical* in nature. This is not in any sense a failing, since there are many situations in research, engineering, the natural sciences, the physical sciences, life science, behavioral science, and other areas in which such empirical knowledge has practical utility or opens new, useful lines of inquiry.

However, along with this diversity of contexts comes a diversity of data. No matter what its intrinsic beauty, a useful model must be flexible enough to adequately support the more specific objectives of *prediction from* or *explanation of* the data presented to it. As we shall see, one of the appealing aspects of partial least squares as a modeling approach is that, unlike some more traditional approaches that might be familiar to you, it is able to encompass much of this diversity within a single framework.

A final comment on modeling in general—*all data is contextual*. Only you can determine the plausibility and relevance of the data that you have, and you overlook this simple fact at your peril. Although statistical modeling can be invaluable, just looking at the data in the right way can and should illuminate and guide the specifics of building empirical statistical models of any kind (Chatfield 1995).

Partial Least Squares in Today's World

Increasingly, we are finding data everywhere. This data explosion, supported by innovative and convergent technologies, has arguably made data exploration (e-Science) a fourth learning paradigm, joining theory, experimentation, and simulation as a way to drive new understanding (Microsoft Research 2009).

In simple retail businesses, sellers and buyers are wrestling for more leverage over the selling/buying process, and are attempting to make better use of data in this struggle. Laboratories, production lines, and even cars are increasingly equipped with relatively low-cost instrumentation routinely producing data of a volume and complexity that was difficult to foresee even thirty years ago. This book shows you how partial least squares, with its appealing flexibility, fits into this exciting picture.

This abundance of data, supported by the widespread use of automated test equipment, results in data sets with a large number of columns, or variables, v and/or a large number of observations, or rows, n. Often, but not always, it is cheap to increase v and expensive to increase n.

When the interpretation of the data permits a natural separation of variables into predictors and responses, partial least squares, or *PLS* for short, is a flexible approach to building statistical models for prediction. PLS can deal effectively with the following:

- Wide data (when $v \gg n$, and v is large or very large)
- Tall data (when $n \gg v$, and n is large or very large)
- Square data (when $n \sim v$, and n is large or very large)
- Collinear variables, namely, variables that convey the same, or nearly the same, information
- Noisy data

Just to whet your appetite, we point out that PLS routinely finds application in the following disciplines as a way of taming multivariate data:

- Psychology
- Education
- Economics
- Political science
- Environmental science
- Marketing
- Engineering
- Chemistry (organic, analytical, medical, and computational)
- Bioinformatics
- Ecology
- Biology
- Manufacturing

Transforming, and Centering and Scaling Data

Data should always be screened for outliers and anomalies prior to any formal analysis, and PLS is no exception. In fact, PLS works best when the variables involved have somewhat symmetric distributions. For that reason, for example, highly skewed variables are often logarithmically transformed prior to any analysis.

Also, the data are usually centered and scaled prior to conducting the PLS analysis. By *centering*, we mean that, for each variable, the mean of all its observations is subtracted from each observation. By *scaling*, we mean that each observation is divided by the variable's standard deviation. Centering and scaling each variable results in a working data table where each variable has mean 0 and standard deviation 1.

The reason that centering and scaling are important is because the weights that form the basis for the PLS model are very sensitive to the measurement units of the variables. Without centering and scaling, variables with higher variance have more influence on the model. The process of centering and scaling puts all variables on an equal footing. If certain variables in **X** are indeed more important than others, and you want them to have higher influence, you can accomplish this by assigning them a higher scaling weight (Eriksson et al. 2006). As you will see, JMP makes centering and scaling easy.

Later we discuss how PLS relates to other modeling and multivariate methods. But for now, let's dive into an example so that we can compare and contrast it to the more familiar multivariate linear regression (MLR).

An Example of a PLS Analysis

The Data and the Goal

The data table Spearheads.jmp contains data relating to the chemical composition of spearheads known to originate from one of two African tribes (Figure 1.1). You can open this table by clicking on the correct link in the master journal. A total of 19 spearheads of known origin were studied. The Tribe of origin is recorded in the first column ("Tribe A" or "Tribe B"). Chemical measurements of 10 properties were made. These are given in the subsequent columns and are represented in the **Columns** panel in a column group called Xs. There is a final column called Set, indicating whether an observation will be used in building our model ("Training") or in assessing that model ("Test").

Figure 1.1: The Spearheads.jmp Data Table

	Tribe	Fe	Ti	Ba	Ca	K	Mn	Rb	Sr	Y	Zr	Set
1	Tribe B	1100	390	55	920	460	45	120	57	58	142	Training
2	Tribe B	1164	404	56	916	446	42	120	58	45	148	Training
3	Tribe B	1030	373	59	920	487	38	128	53	58	138	Training
4	Tribe B	1077	373	55	888	455	38	97	51	54	145	Training
5	Tribe B	1020	360	59	883	473	43	119	40	50	134	Training
6	Tribe B	1100	373	53	910	477	51	137	61	58	152	Training
7	Tribe B	1069	375	51	958	429	42	100	51	47	128	Training
8	Tribe A	1186	257	10	940	431	40	121	20	53	73	Training
9	Tribe A	860	182	7	722	418	33	115	20	55	53	Training
10	Tribe B	1173	417	54	961	441	47	135	55	60	145	Test
11	Tribe B	1080	403	53	919	442	41	133	60	45	155	Test
12	Tribe B	1050	396	56	924	482	48	140	74	71	157	Test
13	Tribe A	863	183	8	626	452	34	121	15	58	70	Test
14	Tribe A	1108	289	7	783	426	41	109	15	57	67	Test
15	Tribe A	1210	276	10	966	430	44	117	20	44	73	Test
16	Tribe A	1205	291	10	975	420	43	115	25	58	73	Test
17	Tribe A	1100	267	10	910	500	40	145	25	65	95	Test
18	Tribe A	1100	280	10	872	515	49	145	38	60	65	Test
19	Tribe A	689	114	9	534	404	26	110	25	50	55	Test

Our goal is to build a model that uses the chemical measurements to help us decide whether other spearheads collected in the vicinity were made by "Tribe A" or "Tribe B". Note that there are 10 columns in **X** (the chemical compositions) and only one column in **Y** (the attribution of the tribe).

The model will be built using the *training set*, rows 1–9. The *test set*, rows 10–19, enables us to assess the ability of the model to predict the tribe of origin for newly discovered spearheads. The column Tribe actually contains the numerical values +1 and −1, with −1 representing "Tribe A" and +1 representing "Tribe B". The Tribe column displays **Value Labels** for these numerical values. It is the numerical values that the model actually predicts from the chemical measurements.

The table Spearheads.jmp also contains four scripts that help us perform the PLS analysis quickly. In the later chapters containing examples, we walk through the menu options that enable you to conduct such an analysis. But, for now, the scripts expedite the analysis, permitting us to focus on the concepts underlying a PLS analysis.

The Analysis

The first script, Fit Model Launch Window, located in the upper left of the data table as shown in Figure 1.2, enables us to set up the analysis we want. From the red-triangle menu, shown in Figure 1.2, select **Run Script**. This script only runs if you are using JMP Pro since it uses the Fit Model partial least squares personality. If you are using JMP, you can select **Analyze > Multivariate Methods > Partial Least Squares** from the JMP menu bar. You will be able to follow the text, but with minor modifications.

Figure 1.2: Running the Script "Fit Model Launch Window"

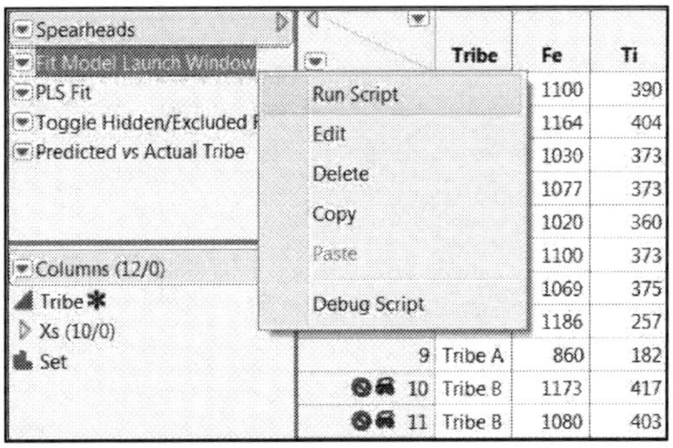

This script produces a populated Fit Model launch window (Figure 1.3). The column Tribe is entered as a response, **Y**, while the 10 columns representing metal composition measurements are entered as **Model Effects**. Note that the **Personality** is set to **Partial Least Squares**. In JMP Pro, you can access this launch window directly by selecting **Analyze > Fit Model** from the JMP menu bar.

Below the **Personality** drop-down menu, shown in Figure 1.3, there are check boxes for **Centering** and **Scaling**. As mentioned in the previous section, centering and scaling all variables in a PLS analysis treats them equitably in the analysis. There is also a check box for **Standardize X**. This option, described in "The Standardize X Option" in Appendix 1, centers and scales columns that are involved in higher-order terms. JMP selects these three options by default.

Figure 1.3: Populated Fit Model Launch Window

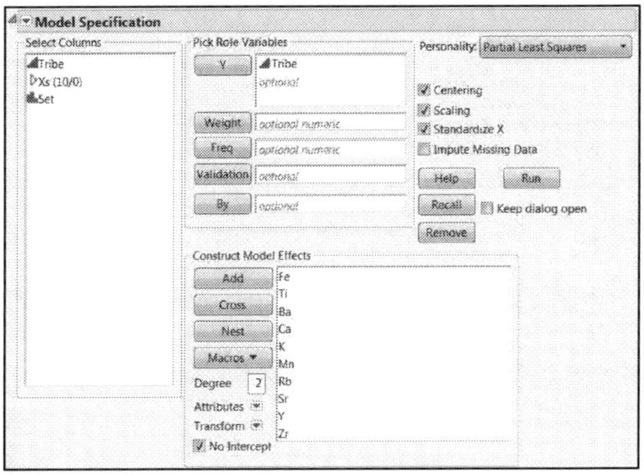

Clicking **Run** brings us to the Partial Least Squares Model Launch control panel (Figure 1.4). Here, we can make choices about how we would like to fit the model. Note that we are allowed to choose between two fitting algorithms to be discussed later: **NIPALS** and **SIMPLS**. We accept the default settings. (To reproduce the exact analysis shown below, select **Set Random Seed** from the red triangle menu at the top of the report and enter 111.) Click **Go**. (You can, instead, run the script PLS Fit to see the report.)

Figure 1.4: PLS Model Launch Control Panel

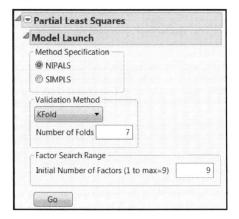

This appends three new report sections, as shown in Figure 1.5: **Model Comparison Summary**, **KFold Cross Validation with K=7 and Method=NIPALS**, and **NIPALS Fit with 3 Factors**. Later, we fully explain the various options and report contents, but for now we take the analysis on trust in order to quickly see this example in its entirety. As we discuss later, the **Number of Factors** is a key aspect of a PLS model. The report in Figure 1.5 shows **3 Factors**, but your report might show a different number. This is because the **Validation Method** of **KFold**, set as a default in the JMP Pro Model Launch control panel, involves an element of randomness.

Figure 1.5: Initial PLS Reports

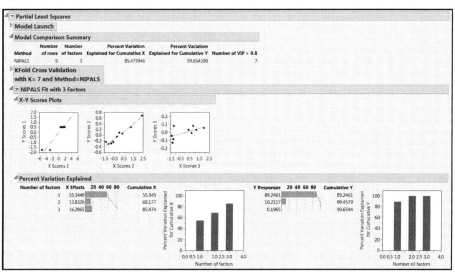

Once you have built a model in JMP, you can save the prediction formula to the table containing the data that were analyzed. We do this for our PLS model. From the options in the red-triangle menu for the **NIPALS Fit with 3 Factors**, select **Save Columns > Save Prediction Formula** (Figure 1.6).

Figure 1.6: Saving the Prediction Formula

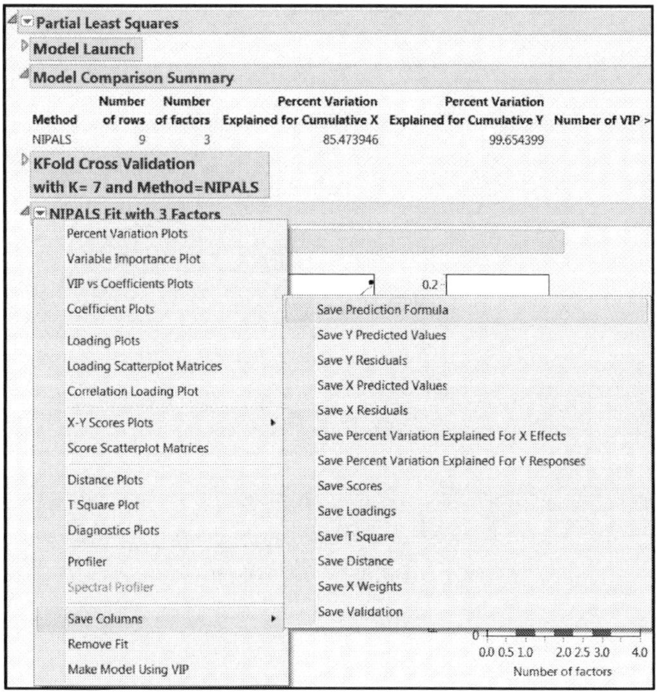

The saved formula column, Pred Formula Tribe, appears as the last column in the data table. Because we are actually saving a formula, we obtain predicted values for all 19 rows.

Testing the Model

To see how well our PLS model has performed, let's simulate the arrival of new data using our test set. We would like to remove the **Hide** and **Exclude** row states from rows 10-19, and apply them to rows 1-9. You can do this by hand, or by running the script Toggle Hidden/Excluded Rows. To do this by hand, select **Rows > Clear Row States**, select rows 1-9, right-click in the highlighted area near the row numbers, and select **Hide and Exclude**. (In versions of JMP prior to JMP 11, select **Exclude/Unexclude**, and then right-click again and select **Hide/Unhide**.)

Now run the script Predicted vs Actual Tribe. For each row, this plots the predicted score for tribal origin on the vertical axis against the actual tribe of origin on the horizontal axis (Figure 1.7).

Figure 1.7: Predicted versus Actual Tribe for Test Data

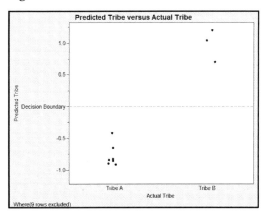

To produce this plot yourself, select **Graph > Graph Builder**. In the **Variables** panel, right-click on the modeling type icon to the left of Tribe and select **Nominal**. (This causes the value labels for Tribe to display.) Drag Tribe to the **X** area and Pred Formula Tribe to the **Y** area.

Note that the predicted values are not exactly +1 or -1, so it makes sense to use a decision boundary (the dotted blue line at the value 0) to separate or *classify* the scores produced by our model into two groups. You can insert a decision boundary by double-clicking on the vertical axis. This opens the Y Axis Specification window. In the **Reference Lines** section near the bottom of the window, click **Add** to add a reference line at 0, and then enter the text **Decision Boundary** in the **Label** text box.

The important finding conveyed by the graph is that our PLS model has performed admirably. The model has correctly classified all ten observations in the test set. All of the observations for "Tribe A" have predicted values below 0 and all those for "Tribe B" have predicted values above 0.

Our model for the spearhead data was built using only nine spearheads, one less than the number of chemical measurements made. PLS provides an excellent classification model in this case.

Before exploring PLS in more detail, let's engage in a quick review of multiple linear regression. This is a common approach to modeling a single variable in **Y** using a collection of variables, **X**.

A Review of Multiple Linear Regression

The Cars Example 11
Estimating the Coefficients 15
Underfitting and Overfitting: A Simulation 16
The Effect of Correlation among Predictors: A Simulation 18

The Cars Example

Consider Figure 2.1, which displays the data table CarsSmall.jmp. You can open this table by clicking on the correct link in the master journal. This data table consists of six rows, corresponding to specific cars of different types, and six variables from the JMP sample data table Cars.jmp.

Figure 2.1: Data Table CarsSmall.jmp

	Automobile	MPG	Number of Cylinders	HP	Weight	Transmission
1	Mazda RX-4	21.0	6	110	2.62	Man
2	Datsun 710	22.8	4	93	2.32	Man
3	Hornet Sportabout	18.7	8	175	3.44	Auto
4	Valiant	18.1	6	105	3.46	Auto
5	Duster 360	14.3	8	245	3.57	Auto
6	Mercedes 240D	24.4	4	62	3.19	Auto

The first column, Automobile, is an identifier column. Our goal is to predict miles per gallon (MPG) from the other descriptive variables. So, in this context, the variable MPG is the single variable in **Y**, and **X** consists of the four variables Number of Cylinders, HP (horsepower), Weight, and Transmission (with values "Man" and "Auto", for manual and automatic transmissions, respectively).

This data structure is typical of the type of data to which multiple linear regression (MLR), or more generally, any modeling approach, is applied. This familiar tabular structure leads naturally to the representation and manipulation of data values as matrices.

To be more specific, a multiple linear regression model for our data can be represented as shown here:

$$(2.1) \quad \begin{pmatrix} 21.0 \\ 22.8 \\ 18.7 \\ 18.1 \\ 14.3 \\ 24.4 \end{pmatrix} = \begin{bmatrix} 1 & 6 & 110 & 2.62 & \text{Man}(0) \\ 1 & 4 & 93 & 2.32 & \text{Man}(0) \\ 1 & 8 & 175 & 3.44 & \text{Auto}(1) \\ 1 & 6 & 105 & 3.46 & \text{Auto}(1) \\ 1 & 8 & 245 & 3.57 & \text{Auto}(1) \\ 1 & 4 & 62 & 3.19 & \text{Auto}(1) \end{bmatrix} * \begin{pmatrix} \beta_0 \\ \beta_1 \\ \beta_2 \\ \beta_3 \\ \beta_4 \end{pmatrix} + \begin{pmatrix} \varepsilon_1 \\ \varepsilon_2 \\ \varepsilon_3 \\ \varepsilon_4 \\ \varepsilon_5 \\ \varepsilon_6 \end{pmatrix}$$

Here are various items to note:

1. The values of the response, MPG, are presented on the left side of the equality sign, in the form of a *column vector*, which is a special type of matrix that contains only a single column. In our example, this is the only column in the response matrix **Y**.

2. The rectangular array to the immediate right of the equality sign, delineated by square brackets, consists of five columns. There is a column of ones followed by four columns consisting of the values of our four predictors, Number of Cylinders, HP (horsepower), Weight, and Transmission. These five columns are the columns in the *matrix* **X**.

3. In parentheses, next to the entries in the last column of **X**, the Transmission value labels, "Man" and "Auto" have been assigned the numerical values 0 and 1, respectively. Because matrices can contain only numeric data, the values of the variable Transmission have to be coded in a numerical form. When a nominal variable is included in a regression model, JMP automatically codes that column, and you can interpret reports without ever knowing what has happened behind the scenes. But if you are curious, select **Help > Books > Fitting Linear Models**, and search for "Nominal Effects" and "Nominal Factors".

4. The column vector consisting of βs, denoted $\boldsymbol{\beta}$, contains the unknown *coefficients* that relate the entries in **X** to the entries in **Y**. These are usually called regression *parameters*.

5. The column vector consisting of epsilons (ε_i), denoted $\boldsymbol{\varepsilon}$, contains the unknown errors. This vector represents the variation that is unexplained when we model **Y** using **X**.

The symbol "$*$" in Equation (2.1) denotes matrix multiplication. The expanded version of Equation (2.1) is:

$$(2.2) \quad \begin{matrix} 21.0 \\ 22.8 \\ 18.7 \\ 18.1 \\ 14.3 \\ 24.4 \end{matrix} = \begin{matrix} \beta_0 + 6\beta_1 + 110\beta_2 + 2.62\beta_3 + \varepsilon_1 \\ \beta_0 + 4\beta_1 + 93\beta_2 + 2.32\beta_3 + \varepsilon_2 \\ \beta_0 + 8\beta_1 + 175\beta_2 + 3.44\beta_3 + \beta_4 + \varepsilon_3 \\ \beta_0 + 6\beta_1 + 105\beta_2 + 3.46\beta_3 + \beta_4 + \varepsilon_4 \\ \beta_0 + 8\beta_1 + 245\beta_2 + 3.57\beta_3 + \beta_4 + \varepsilon_5 \\ \beta_0 + 4\beta_1 + 62\beta_2 + 3.19\beta_3 + \beta_4 + \varepsilon_6 \end{matrix}$$

Equation (2.2) indicates that each response is to be modeled as a linear function of the unknown βs.

We can represent Equation (2.1) more generically as:

$$(2.3) \quad \begin{bmatrix} Y_1 \\ Y_2 \\ Y_3 \\ Y_4 \\ Y_5 \\ Y_6 \end{bmatrix} = \begin{bmatrix} X_{10} & X_{11} & X_{12} & X_{13} & X_{14} \\ X_{20} & X_{21} & X_{22} & X_{23} & X_{24} \\ X_{30} & X_{31} & X_{32} & X_{33} & X_{34} \\ X_{40} & X_{41} & X_{42} & X_{43} & X_{44} \\ X_{50} & X_{51} & X_{52} & X_{53} & X_{54} \\ X_{60} & X_{61} & X_{62} & X_{63} & X_{64} \end{bmatrix} * \begin{bmatrix} \beta_0 \\ \beta_1 \\ \beta_2 \\ \beta_3 \\ \beta_4 \end{bmatrix} + \begin{bmatrix} \varepsilon_1 \\ \varepsilon_2 \\ \varepsilon_3 \\ \varepsilon_4 \\ \varepsilon_5 \\ \varepsilon_6 \end{bmatrix}$$

Now we can write Equation (2.3) succinctly as:

$$(2.4) \quad \mathbf{Y} = \mathbf{X}\boldsymbol{\beta} + \boldsymbol{\varepsilon},$$

Here

$$\mathbf{Y} = \begin{bmatrix} Y_1 \\ Y_2 \\ Y_3 \\ Y_4 \\ Y_5 \\ Y_6 \end{bmatrix},$$

$$\mathbf{X} = \begin{matrix} X_{10} & X_{11} & X_{12} & X_{13} & X_{14} \\ X_{20} & X_{21} & X_{22} & X_{23} & X_{24} \\ X_{30} & X_{31} & X_{32} & X_{33} & X_{34} \\ X_{40} & X_{41} & X_{42} & X_{43} & X_{44} \\ X_{50} & X_{51} & X_{52} & X_{53} & X_{54} \\ X_{60} & X_{61} & X_{62} & X_{63} & X_{64} \end{matrix},$$

$$\boldsymbol{\beta} = \begin{matrix} \beta_0 \\ \beta_1 \\ \beta_2 \\ \beta_3 \\ \beta_4 \end{matrix},$$

and

$$\boldsymbol{\varepsilon} = \begin{matrix} 1 \\ 2 \\ 3 \\ 4 \\ 5 \\ 6 \end{matrix}$$

For a column vector like **Y**, we need only one index to designate the row in which an element occurs. For the 6 by 5 matrix **X**, we require two indices. The first designates the row and the second designates the column. Note that we have not specified the matrix multiplication operator in Equation (2.4); it is implied by the juxtaposition of any two matrices.

Equation (2.4) enables us to note the following:

1. The entries in **X** consist of the column of ones followed by the observed data on each of the four predictors.

2. Even though the entries in **X** are observational data, rather than the result of a designed experiment, the matrix **X** is still called the *design matrix*.

3. The vector $\boldsymbol{\varepsilon}$, which contains the errors, $_i$, is often referred to as the *noise*.

4. Once we have estimated the column vector $\boldsymbol{\beta}$, we are able to obtain predicted values of MPG. By comparing these predicted values to their actual values, we

obtain estimates of the errors, ε_i. These differences, namely the actual minus the predicted values, are called *residuals*.

5. If the model provides a good fit, we expect the residuals to be small, in some sense. We also expect them to show a somewhat random pattern, indicating that our model adequately captures the structural relationship between **X** and **Y**. If the residuals show a structured pattern, one remedy might be to specify a more complex model by adding additional columns to **X**; for example, columns that define *interaction* terms and/or *power* terms (Draper and Smith 1998).

6. The values in **β** are the *coefficients* or *parameters* that correspond to each column or *term* in the design matrix **X** (including the first, constant term). In terms of this linear model, their interpretation is straightforward. For example, β_3 is the expected change in MPG for a unit change in Weight.

7. Note that the *dimensions* of the matrices (number of rows and columns) have to conform in Equation (2.4). In our example, **Y** is a 6 by 1 matrix, **X** is a 6 by 5 matrix, **β** is a 5 by 1 matrix, and **ε** is a 6 by 1 matrix.

Estimating the Coefficients

So how do we calculate **β** from the data we have collected? There are numerous approaches, partly depending on the assumptions you are prepared or required to make about the noise component, **ε**. It is generally assumed that the **X** variables are measured without noise, so that the noise is associated only with the measurement of the response, **Y**.

It is also generally assumed that the errors, ε_i, are identically and independently distributed according to a normal distribution (Draper and Smith 1998). Once a model is fit to the data, your next step should be to check if the pattern of residuals is consistent with this assumption.

For a full introduction to MLR in JMP using the Fit Model platform, select **Help > Books > Fitting Linear Models**. When the PDF opens, go to the chapter entitled "Standard Least Squares Report and Options."

More generally, returning to the point about matrix dimensions, the dimensions of the components of a regression model of the form

(2.5) $$\mathbf{Y} = \mathbf{X}\boldsymbol{\beta} + \boldsymbol{\varepsilon}$$

can be represented as follows:

- **Y** is an $n \times 1$ response matrix.
- **X** is an $n \times m$ design matrix
- **β** is an $m \times 1$ coefficient vector.
- **ε** is an $n \times 1$ error vector.

Here n is the number of observations and m is the number of columns in **X**. For now, we assume that there is only one column in **Y**, but later on, we consider situations where **Y** has multiple columns.

Let's pause for a quick linear algebra review. If **A** is any $r \times s$ matrix with elements (a_{ij}), then the matrix **A'**, with elements (a_{ji}), is called the *transpose* of **A**. Note that the rows of **A** are the columns of **A'**. We denote the inverse of a square $q \times q$ matrix **B** by $\mathbf{B}^{-1}$. By definition, $\mathbf{BB}^{-1} = \mathbf{B}^{-1}\mathbf{B} = \mathbf{I}$, where **I** is the $q \times q$ identity matrix (with 1's down the leading diagonal and 0's elsewhere). If the columns of an arbitrary matrix, say **A**, are linearly independent, then it can be shown that the inverse of the matrix **A'A** exists.

In MLR, when $n \geq m$ and when the columns of **X** are linearly independent so that the matrix $(\mathbf{X'X})^{-1}$ exists, the coefficients in **β** can be estimated in a unique fashion as:

$$\hat{\boldsymbol{\beta}} = (\mathbf{X'X})^{-1}\mathbf{X'Y}$$

The hat above **β** in the notation $\hat{\boldsymbol{\beta}}$ indicates that this vector contains numerical estimates of the unknown coefficients in **β**. If there are fewer observations than columns in **X**, $n < m$, then there are an infinite number of solutions for **β** in Equation (2.5).

As an example, think of trying to fit two observations with a matrix **X** that has three columns. Then, geometrically, the expression **Xβ** in Equation (2.5) defines a hyperplane which, given that $m = 3$ in this case, is simply a plane. But there are infinitely many planes that pass through any two given points. There is no way to determine which of these infinitely many solutions would be best at predicting new observations well.

Underfitting and Overfitting: A Simulation

To better understand the issues behind model fitting, let's run the script **PolyRegr.jsl** by clicking on the correct link in the master journal.

The script randomly generates **Y** values for eleven points with **X** values plotted horizontally and equally spaced over the interval 0 to 1 (Figure 2.2). The points exhibit some curvature. The script uses MLR to predict **Y** from **X**, using various polynomial models. (Note that your points will differ from ours because of the randomness.)

When the slider in the bottom left corner is set at the far left, the order of the polynomial model is one. In other words, we are fitting the data with a line. In this case, the design matrix **X** has two columns, the first containing all 1s and the second containing the horizontal coordinates of the plotted points. The linear fit ignores the seemingly obvious pattern in the data— it is *underfitting* the data. This is evidenced by the residuals, whose magnitudes are illustrated using vertical blue lines. The RMSE (*root mean square error*) is calculated by squaring each residual, averaging these (specifically, dividing their sum by the number of observations minus one, minus the number of predictors), and then taking the square root.

As we shift the slider to the right, we are adding higher-order polynomial terms to the model. This is equivalent to adding additional columns to the design matrix. The additional polynomial terms provide a more flexible model that is better able to capture the important characteristics, or the *structure*, of the data.

Figure 2.2 Illustration of Underfitting and Overfitting, with Order = 1, 2, 3, and 10

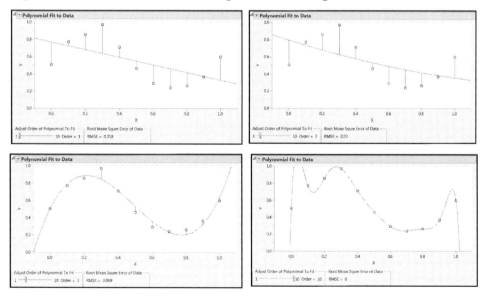

However, we get to a point where we go beyond modeling the structure of the data, and begin to model the noise in the data. Note that, as we increase the order of the polynomial, thereby adding more terms to the model, the RMSE progressively reduces. An order 10 polynomial, obtained by setting the slider all the way to the right, provides a perfect fit to the data and gives RMSE = 0 (bottom right plot in Figure 2.2). However, this model is not generalizable to new data, because it has modeled both the structure and the noise, and by definition the noise is random and unpredictable. Our model has *overfit* the data.

In fitting models, we must strike a balance between modeling the intrinsic structure of the data and modeling the noise in the data. One strategy for reaching this goal is the use of *cross-validation*, which we shall discuss in the section "Choosing the Number of Factors" in Chapter 4. You can close the report produced by PolyRegr.jsl at this point.

The Effect of Correlation among Predictors: A Simulation

In MLR, correlation among the predictors is called *multicollinearity*. We explore the effect of multicollinearity on estimates of the regression coefficients by running the script Multicollinearity.jsl. Do this by clicking on the correct link in the master journal. The script produces the launch window shown in Figure 2.3.

Figure 2.3: Multicollinearity Simulation Launch Window

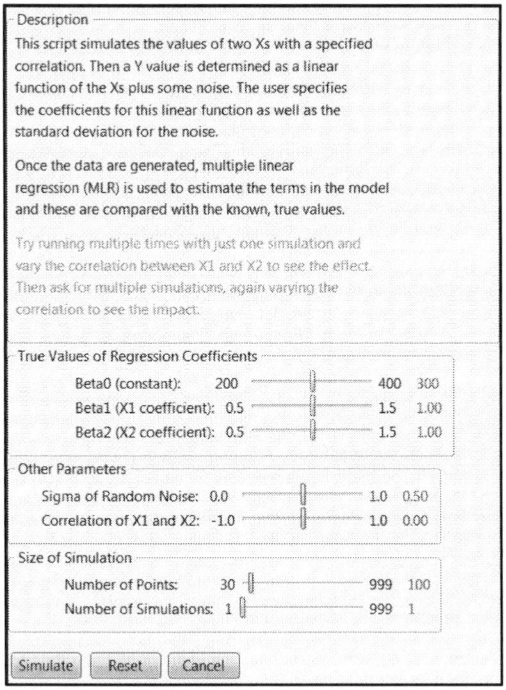

Description

This script simulates the values of two Xs with a specified correlation. Then a Y value is determined as a linear function of the Xs plus some noise. The user specifies the coefficients for this linear function as well as the standard deviation for the noise.

Once the data are generated, multiple linear regression (MLR) is used to estimate the terms in the model and these are compared with the known, true values.

Try running multiple times with just one simulation and vary the correlation between X1 and X2 to see the effect. Then ask for multiple simulations, again varying the correlation to see the impact.

True Values of Regression Coefficients

Beta0 (constant):	200		400	300
Beta1 (X1 coefficient):	0.5		1.5	1.00
Beta2 (X2 coefficient):	0.5		1.5	1.00

Other Parameters

Sigma of Random Noise:	0.0		1.0	0.50
Correlation of X1 and X2:	-1.0		1.0	0.00

Size of Simulation

Number of Points:	30		999	100
Number of Simulations:	1		999	1

[Simulate] [Reset] [Cancel]

The launch window enables you to set conditions to simulate data from a known model:

- You can set the values of the three regression coefficients: **Beta0 (constant)**; **Beta1 (X1 coefficient)**; and **Beta2 (X2 coefficient)**. Because there are three regression parameters, you are defining a plane that models the mean of the response, **Y**. In symbols,

$$E[Y] = \beta_0 + \beta_1 X1 + \beta_2 X2$$

where the notation $E[Y]$ represents the expected value of **Y**.

- The noise that is applied to **Y** is generated from a normal distribution with mean 0 and with the standard deviation that you set as **Sigma of Random Noise** under **Other Parameters**. In symbols, this means that ε in the expression

$$Y = \beta_0 + \beta_1 X1 + \beta_2 X2 + \varepsilon$$

has a normal distribution with mean 0 and standard deviation equal to the value you set.

- You can specify the correlation between the values of X1 and X2 using the slider for **Correlation of X1 and X2** under **Other Parameters**. X1 and X2 values will be generated for each simulation from a multivariate normal distribution with the specified correlation.
- In the **Size of Simulation** panel, you can specify the **Number of Points** to be generated for each simulation, as well as the **Number of Simulations** to run.

Once you have set values for the simulation using the slider bars, generate results by clicking **Simulate**. Depending on your screen size, you can view multiple results simultaneously without closing the launch window.

Let's first run a simulation with the initial settings. Then run a second simulation after moving the **Correlation of X1 and X2** slider to a large, positive value. (We have selected 0.92.) Your reports will be similar to those shown in Figure 2.4.

Figure 2.4: Comparison of Design Settings, Low and High Predictor Correlation

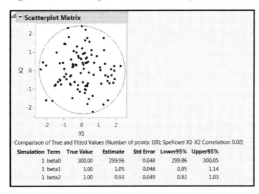

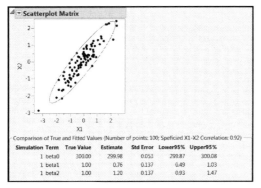

The two graphs reflect the differences in the settings of the **X** variables for the two correlation scenarios. In the first, the points are evenly distributed in a circular pattern. In the second, the points are condensed into a narrow elliptical pattern. These patterns show the geometry of the design matrix for each scenario.

In the high correlation scenario, note that high values of X1 tend to be associated with high values of X2, and that low values of X1 tend to be associated with low values of X2. This is exactly what is expected for positive correlation. (For a definition of the correlation coefficient between observations, select **Help > Books > Multivariate Methods** and search for "Pearson Product-Moment Correlation").

The true and estimated coefficient values are shown at the bottom of each plot. Because our model was not deterministic—the **Y** values were generated so that their means are linear functions of X1 and X2, but the actual values are affected by noise—the estimated coefficients are just that, estimates, and as such, they reflect uncertainty. This uncertainty is quantified in the columns **Std Error** (standard error), **Lower95%** (the lower 95% limit for the coefficient's confidence interval), and **Upper95%** (the upper 95% limit for the coefficient's confidence interval).

Notice that the estimates of **beta1** and **beta2** can be quite different from the true values in the high correlation scenario (bottom plot in Figure 2.4). Consistent with this, the standard errors are larger and the confidence intervals are wider.

Let's get more insight on the impact that changing the correlation value has on the estimates of the coefficients. Increase the **Number of Simulations** to about 500 using the slider, and again simulate with two different values of the correlation, one near zero and one near one. You should obtain results similar to those in Figure 2.5.

Figure 2.5: Plots of Estimates for Coefficients, Low and High Predictor Correlation

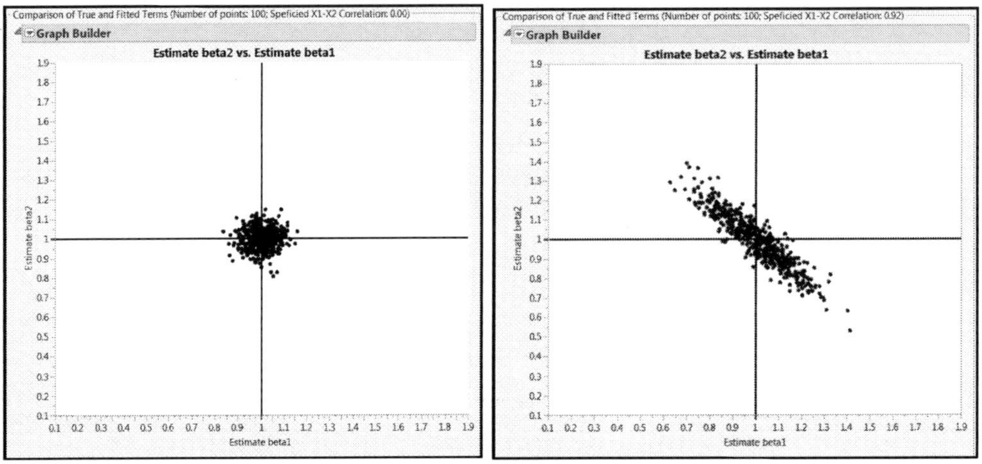

These plots show **Estimate beta1** and **Estimate beta2**, the estimated values of β_1 and β_2, from the 500 or so regression fits. The reference lines show the true values of the corresponding parameters, so that the intersection of these two lines shows the pair of true values used to simulate the data. In an ideal world, all of the estimate pairs would be very close to the point defined by the true values.

When X1 and X2 have correlation close to zero, the parameter estimates cluster rather uniformly around the true value. However, the impact of high correlation between X1 and X2 is quite dramatic. As this correlation increases, the estimates of β_1 and β_2 become much more variable, but also more strongly (and negatively) correlated themselves. As the correlation between the two predictors X1 and X2 approaches +1.0 or –1.0, we say that the $\mathbf{X'X}$ matrix involved in the MLR solution, becomes *ill-conditioned* (Belsley 1991).

In fact, when there is perfect correlation between X1 and X2, the MLR coefficient estimates cannot be computed because the matrix $(\mathbf{X'X})^{-1}$ does not exist. The situation is similar to trying to build a regression model for MPG with two redundant variables, say, "Weight of Car in Kilograms" and "Weight of Car in Pounds." Because the two predictors are redundant, there really is only a single predictor, and the MLR algorithm doesn't know where to place its coefficients. There are infinitely many ways that the coefficients can be allocated to both terms to produce the same model.

In cases of multicollinearity, the coefficient estimates are highly variable, as you see in Figure 2.5. This means that estimates have high standard errors, so that confidence intervals for the parameters are wide. Also, hypothesis tests can be ineffective because of the uncertainty inherent in the parameter estimates. Much research has been devoted to detecting multicollinearity and dealing with its consequences. Ridge regression and the lasso method (Hastie et al. 2001) are examples of regularization techniques that can be useful when high multicollinearity is present. (In JMP Pro 11, select **Help > Books > Fitting Linear Models** and search for "Generalized Regression Models".)

Whether multicollinearity is of concern depends on your modeling objective. Are you interested in explaining or predicting? Multicollinearity is more troublesome for explanatory models, where the goal is to figure out which predictors have an important effect on the response. This is because the parameter estimates have high variability, which negatively impacts any inference about the predictors. For prediction, the model is useful, subject to the general caveat that an empirical statistical model is good only for interpolation, rather than extrapolation. For example, in the correlated case shown in Figure 2.4, one would not be confident in making predictions when X1 = +1 and X2 = -1 because the model is not supported by any data in that region.

You can close the reports produced by Multicollinearity.jsl at this point.

Principal Components Analysis: A Brief Visit

Principal Components Analysis 25
Centering and Scaling: An Example 28
The Importance of Exploratory Data Analysis in Multivariate Studies 31
Dimensionality Reduction via PCA 34

Principal Components Analysis

Like PLS, principal components analysis (PCA) attempts to use a relatively small number of components to model the information in a set of data that consists of many variables. Its goal is to describe the internal structure of the data by modeling its variance. It differs from PLS in that it does not interpret variables as inputs or outputs, but rather deals only with a *single* matrix. The single matrix is usually denoted by **X**. Although the components that are extracted can be used in predictive models, in PCA there is no direct connection to a **Y** matrix.

Let's look very briefly at an example. Open the data table Solubility.jmp by clicking on the correct link in the master journal. This JMP sample data table contains data on 72 chemical compounds that were measured for solubility in six different solvents, and is shown in part in Figure 3.1. The first column gives the name of the compound. The next six columns give the solubility measurements. We would like to develop a better understanding of the essential features of this data set, which consists of a 72 x 6 matrix.

Figure 3.1: Partial View of Solubility.jmp

	Labels	1-Octanol	Ether	Chloroform	Benzene	Carbon Tetrachloride	Hexane
1	METHANOL	-0.770	-1.150	-1.260	-1.890	-2.100	-2.800
2	ETHANOL	-0.310	-0.570	-0.850	-1.620	-1.400	-2.100
3	PROPANOL	0.250	-0.020	-0.400	-0.700	-0.820	-1.520
4	BUTANOL	0.880	0.890	0.450	-0.120	-0.400	-0.700
5	PENTANOL	1.560	1.200	1.050	0.620	0.400	-0.400
6	HEXANOL	2.030	1.800	1.690	1.300	0.990	0.460
7	HEPTANOL	2.410	2.400	2.410	1.910	1.670	1.010
8	ACETIC_ACID	-0.170	-0.340	-1.600	-2.260	-2.450	-3.060
9	PROPIONICACID	0.330	0.270	-0.960	-1.350	-1.600	-2.140
10	BUTYRICACID	0.790	0.610	-0.270	-0.960	-0.970	-1.760
11	HEXANOICACID	1.920	1.950	1.150	0.300	0.570	-0.460
12	PENTANOICACID	1.390	1.000	0.280	-0.100	-0.420	-1.000
13	TRICHLOROACETICACID	1.330	1.210	-0.690	-1.300	-1.660	-2.630
14	DICHLOROACETICACID	0.920	1.310	-0.890	-1.400	-2.310	-2.720
15	CHLOROACETICACID	0.220	0.370	-1.920	-1.600	-2.560	-3.140
16	METHYLACETATE	0.180	0.430	1.160	0.530	0.320	-0.260
17	ETHYLACETATE	0.730	0.930	1.800	1.010	0.950	0.290
18	ACETONE	-0.240	-0.210	0.240	-0.050	-0.300	-0.910

PCA works by extracting linear combinations of the variables. First, it finds a linear combination of the variables that maximizes the variance. This is done subject to a constraint on the sizes of the coefficients, so that a solution exists. Subject to this constraint, the first linear combination explains as much of the variability in the data as possible. The observations are then weighted by this linear combination, to produce scores. The vector of scores is called the *first principal component*. The vector of coefficients for the linear combination is sometimes called the first *loading* vector.

Next, PCA finds a linear combination which, among all linear combinations that are orthogonal to the first, has the highest variance. (Again, a constraint is placed on the sizes of the coefficients.) This second vector of factor loadings is used to compute scores for the observations, resulting in the *second principal component*. This second principal component explains as much variance as possible in a direction orthogonal to that of the first loading vector. Subsequent linear combinations are extracted similarly, to explain the maximum variance in the space that is orthogonal to the loading vectors that have been previously extracted.

To perform PCA for this data set in JMP:

1. Select **Analyze > Multivariate Methods > Principal Components**.
2. Select the columns 1-Octanol through Hexane and add them as **Y, Columns**.
3. Click **OK**.

4. In the red triangle menu for the resulting report, select **Eigenvalues**.

Your report should appear as in Figure 3.2. (Alternatively, you can simply run the last script in the data table panel, Principal Components.)

Figure 3.2: PCA Analysis for Solubility.jmp

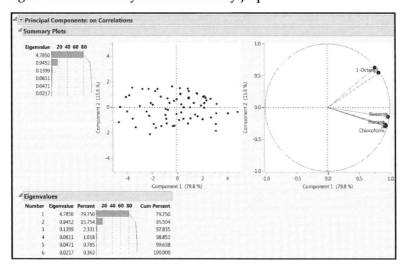

Each row of data is transformed to a score on each principal component. Plots of these scores for the first two principal components are shown. We won't get into the technical details, but each component has an associated eigenvalue and eigenvector. The **Eigenvalues** report indicates that the first component accounts for 79.75% of the variation in the data, and that the second component brings the cumulative total variation accounted for to 95.50%.

The plot on the far right in Figure 3.2, called a *loading plot*, gives insight into the data structure. All six of the variables have positive loadings on the first component. This means that the largest component of the variability is explained by a linear combination of all six variables with positive coefficients for each variable. But the second component has positive loadings only for 1-Octanol and Ether, while all other variables have negative loadings. This indicates that the next largest source of variability results from a difference between a compound's solubility for 1-Octanol and Ether and the other four solvents.

For more information about the PCA platform in JMP, select **Analyze > Multivariate Methods > Principal Components**. Then, in the launch window, click **Help**.

Centering and Scaling: An Example

As mentioned in Chapter 1, multivariate methods, such as PCA, are very sensitive to the scale of the data. Open the data table **LoWarp.jmp** by clicking on the correct link in the master journal. These data, presented in Eriksson et al. (2006), are from an experiment run to minimize the warp and maximize the strength of a polymer used in a mobile phone. In the column group called Y, you see eight measurements of warp (warp1 – warp8) and six measurements of strength (strength1 – strength6).

Run the first script in the table, Raw Y. You obtain a plot of comparative box plots, as shown in Figure 3.3. Most of the box plots are dwarfed compared to the four larger box plots.

Figure 3.3: Comparative Box Plots for Raw Data

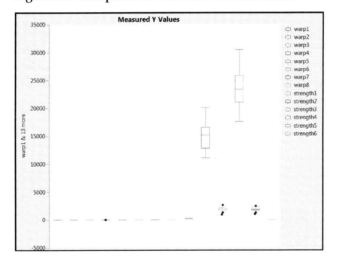

The plot shows that the variables strength2 and strength4 dominate in terms of the raw measurement scale – their values are much larger than those of the other variables. If these raw values were used in a multivariate analysis such as PCA or PLS, these two variables would dominate.

We can lessen the impact of the size of their values by subtracting each variable's mean from all its measurements. As mentioned in Chapter 1, this is called *centering*. The **Columns** panel in the data table contains a column group for the centered variables, called Y Centered. Each variable in this group is given by a formula. To see the formulas, click on the disclosure icon to the left of the group name to display the 13 variables in the group. Next, click on any of the **+** signs to the right of the variable names. You see that the calculated values in any given column consist of the raw data minus the appropriate column mean.

Run the script **Centered Y** to obtain the box plots for the centered data shown in Figure 3.4. Although the data are now centered at 0, the variables strength2_Centered and strength4_Centered still dominate because of their relatively high variability.

Figure 3.4: Comparative Box Plots for Centered Data

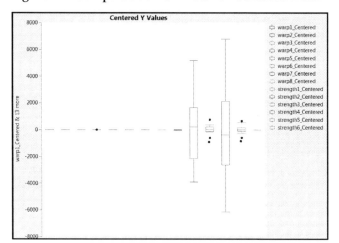

Let's not only center the data in any given column, but let's also divide these centered values by the standard deviation of the column to *scale* the data. JMP has a function that both centers and scales a column of data. The function is called **Standardize**. The column group Y Centered and Scaled contains standardized versions of each of the variables. You can check this by looking at the formulas that define the columns. Run the script Centered and Scaled Y to obtain the comparative box plots of the standardized variables shown in Figure 3.5.

Figure 3.5: Comparative Box Plots for Centered and Scaled Data

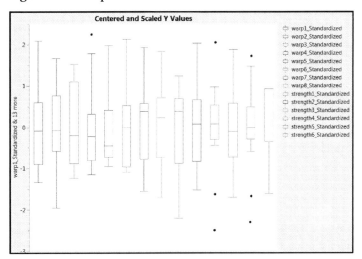

As mentioned in Chapter 1, we see that the act of centering and scaling (or standardizing) the variables does indeed place all of them on an equal footing. Although there can be exceptions, it is generally the case that, in PCA and PLS, centering and scaling your variables is desirable.

In PCA, the JMP default is to calculate **Principal Components > on Correlations**, as shown in Figure 3.6. This means that the variables are first centered and scaled, so that the matrix containing their inner products is the correlation matrix. JMP also enables the user to select **Principal Components > on Covariances**, which means that the data are simply centered, or **Principal Components > on Unscaled**, which means that the raw data are used.

Figure 3.6: PCA Default Calculation

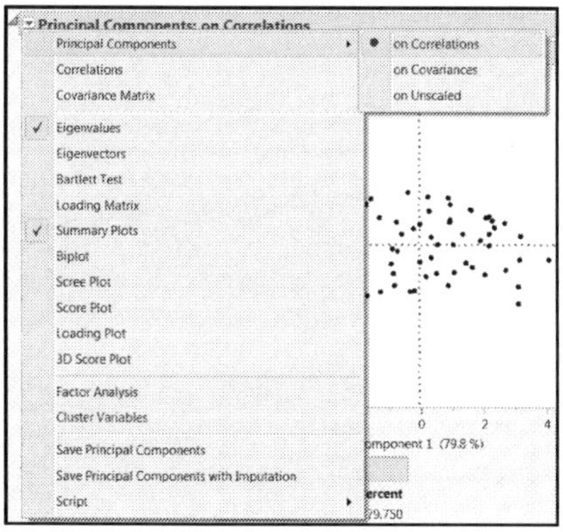

The Importance of Exploratory Data Analysis in Multivariate Studies

Visual data exploration should be a first step in any multivariate study. In the next section, we use some simulated data to see how PCA reduces dimensionality. But first, let's explore the data that we use for that demonstration.

Run the script DimensionalityReduction.jsl by clicking on the correct link in the master journal. This script generates three panels. Each panel gives a plot of 11 quasi-random values for two variables. The first panel that appears when you run the script shows the raw data for X1 and X2, which we refer to as the **Measured Data Values** (Figure 3.7). Your plot will be slightly different, because the points are random, and the **Summary of Measured Data** information will differ to reflect this.

Figure 3.7: Panel 1, Measured Data Values

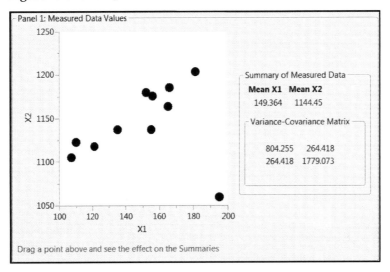

In Panel 1, the **Summary of Measured Data** gives the mean of each variable and the **Variance-Covariance Matrix** for X1 and X2. Note that the variance-covariance matrix is symmetric. The diagonal entries are the variance of X1 (upper left) and the variance of X2 (lower right), while the off-diagonal entries give the covariance of X1 and X2.

Covariance measures the joint variation in X1 and X2. Because the covariance value depends on the units of measurement of X1 and X2, its absolute size is not all that meaningful. However, the pattern of joint variation between X1 and X2 can be discerned and assessed from the scatterplot in Figure 3.7. Note that, as X1 increases, X2 tends to increase as well, but there appears to be one point in the lower right corner of the plot that doesn't fit this general pattern. Although the points generated by the script are random, you should see an outlier in the lower right corner of your plot as well.

Panel 2, shown in Figure 3.8, displays the **Centered and Scaled Data Values**. For the centered and scaled data, the covariance matrix is just the correlation matrix. Here, the off-diagonal entries give the correlation between X1 and X2. This correlation value does have an interpretation based on its size and its sign. In our example, the correlation is 0.221, indicating a weak positive relationship.

Figure 3.8: Panel 2, Centered and Scaled Data Values

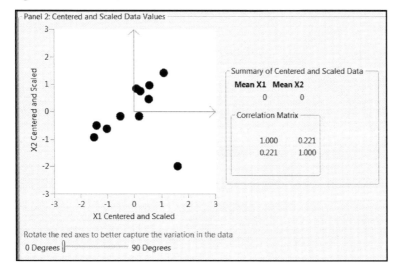

However, as you might suspect, the outlying point might be causing the correlation coefficient to be smaller than expected. In the top panel, you can use your mouse to drag that outlying point, which enables you to investigate its effect on the various numerical summaries, and in particular, on the correlation shown in the second panel. The effect of moving the rogue point into the cloud of the remaining X1, X2 values is shown for our data in Figure 3.9. (The point is now the one at the top right.) The correlation increases from 0.221 to 0.938.

Figure 3.9: Effect of Dragging Outlier to Cloud of Points

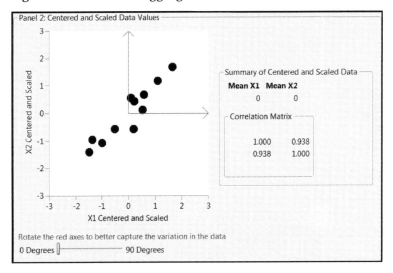

We move on to Panel 3 in the next section, but first a few remarks. Remember that PCA is based on a correlation matrix. We see later that the same is true of PLS. This large change in the correlation highlights the importance of checking your data for incongruous samples before conducting any analysis. In fact, the importance of *exploratory data analysis* (EDA) increases as the number of variables increases. JMP facilitates this task through its interactivity and dynamically linked views.

In Chapter 2, "A Review of Multiple Linear Regression," we saw that fitting a regression model produces residuals. Residuals can also form a basis for detecting incongruous values in multivariate procedures such as PCA and PLS. However, one must remember that residuals only judge samples in relation to the model currently being considered. It is always best to precede model building with exploratory analysis of one's data.

Dimensionality Reduction via PCA

In the context of the data table Solubility.jmp, we saw that the first two principal components explain about 95% of the variation in the six variables. Let's continue using the script DimensionalityReduction.jsl to gain some intuition for exactly how PCA reduces dimensionality.

With the slider in Panel 2 set at **0 Degrees**, we see the axes shown in Figure 3.10. The vertical blue lines in Panel 3 show the distances of the points from the horizontal axis.

The sum of the squares of these distances, called the **Sum of Squared Residuals**, is given to the right of the plot. This sum equals 10 for your simulated data as well as for ours. This is a consequence of the fact that the sum is computed for the centered and scaled data and that there are 11 data points.

Figure 3.10: No Rotation of Axes

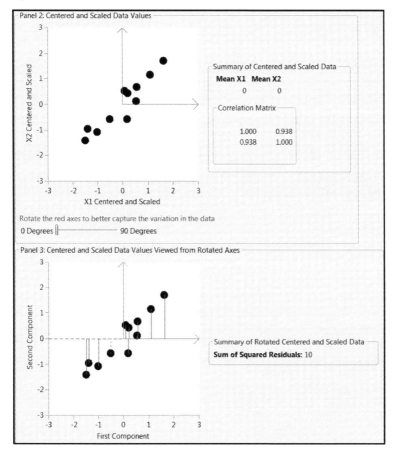

But now, move your slider in Panel 2 to the right to rotate the axes until you come close to minimizing the **Sum of Squared Residuals** given in Panel 3 (Figure 3.11). The total length of the blue lines in the third panel is greatly reduced. In effect, the third panel gives a view of the cloud of data from a rotated coordinate system (defined by the red axes in the second panel).

Figure 3.11: Rotation of Axes

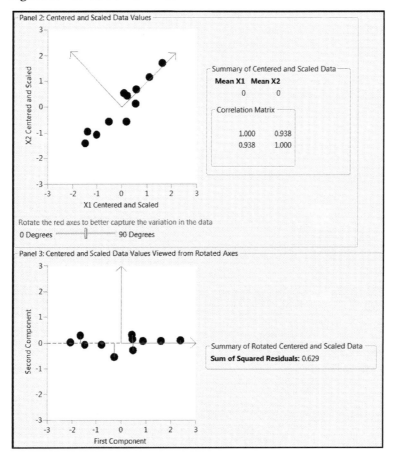

From this new point of view, we have explained much of the variation in the data using a single coordinate. We can think of each point as being projected onto the horizontal line in Panel 3, or, equivalently, onto the rotated axis pointing up and to the right in the second panel. In fact, PCA proceeds in just this manner, identifying the first *principal component* to be the axis along which the variation of the projected points is maximized.

You can close the report generated by the script DimensionalityReduction.jsl.

A Deeper Understanding of PLS

Centering and Scaling in PLS 37
PLS as a Multivariate Technique 38
Why Use PLS? 39
How Does PLS Work? 45
PLS versus PCA 49
PLS Scores and Loadings 50
    Some Technical Background 50
An Example Exploring Prediction 59
    One-Factor NIPALS Model 60
    Two-Factor NIPALS Model 63
    Variable Selection 64
    SIMPLS Fits 64
Choosing the Number of Factors 65
    Cross Validation 65
    Types of Cross Validation 66
    A Simulation of K-Fold Cross Validation 67
    Validation in the PLS Platform 69
The NIPALS and SIMPLS Algorithms 71
Useful Things to Remember About PLS 72

Centering and Scaling in PLS

Although it can be adapted to more general situations, PLS usually involves only two sets of variables, one interpreted as predictors, **X**, and one as responses, **Y**. As with PCA, it is usually best to apply PLS to data that have been centered and scaled. As shown in Chapter 3 this puts all variables on an equal footing. This is why the **Centering** and

Scaling options are turned on by default in the JMP PLS launch window.

There are sometimes cases where it might be useful or necessary to scale blocks of variables in **X** and/or **Y** differently. This can easily be done using JMP column formulas (as we saw in LoWarp.jmp) or using JMP scripting (**Help > Books > Scripting Guide**). In cases where you define your own scaling, be sure to deselect the relevant options in the PLS launch window. For simplicity, we assume for now that we always want all variables to be centered and scaled.

PLS as a Multivariate Technique

When all variables are centered and scaled, their covariance matrix equals their correlation matrix. The correlation matrix becomes the natural vehicle for representing the relationship between the variables. We have already talked about correlation, specifically in the context of predictors and the **X** matrix in MLR.

But the distinction between predictors and responses is contextual. Given any data matrix, we can compute the sample correlation between each pair of columns regardless of the interpretation we choose to assign to the columns. The sample correlations form a square matrix with ones on the main diagonal. *Most linear multivariate methods, PLS included, start from a consideration of the correlation matrix* (Tobias 1995).

Suppose, then, that we have a total of v variables measured on our n units or samples. We consider k of these to be responses and m of these to be predictors, so that $v = k + m$. The correlation matrix, denoted Σ, is $v \times v$. Suppose that $k = 2$ and $m = 4$. Then we can represent the correlation matrix schematically as shown in Figure 4.1, where we order the variables in such a way that responses come first.

Figure 4.1: Schematic of Correlation Matrix Σ, for Ys and Xs

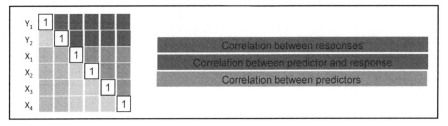

Recall that the elements of Σ must be between –1 and +1 and that the matrix is square and symmetric. But not every matrix with these properties is a correlation matrix. Because of the very meaning of correlation, the elements of a correlation matrix cannot be completely independent of one another. For example, in a 3 x 3 correlation matrix there are three off-diagonal elements: If one of these elements is 0.90 and another is –0.80, it is possible to show that the third must be between –0.98 and –0.46.

As we soon see in a simulation, PLS builds models linking predictors and responses. PLS does this using projection to reduce dimensionality by extracting factors (also called *latent variables* in the PLS context). The information it uses to do this is contained in the dark green elements of **Σ**, in this case a 2 x 4 submatrix. For general values of k and m, this sub-matrix is not square and does not have any special symmetry. We describe exactly how the factors are constructed in Appendix 1.

For now, though, note that the submatrix used by PLS contains the correlations between the predictors and the responses. Using these correlations, the factors are extracted in such a way that they not only explain variation in the **X** and **Y** variables, but they also relate the **X** variables to the **Y** variables.

As you might suspect, consideration of the entire 6 x 6 correlation matrix without regard to predictors and responses leads directly to PCA. As we have seen in Chapter 3, PCA also exploits the idea of projections to reduce dimensionality, and is often used as an exploratory technique prior to PLS.

Consideration of the 4 x 4 submatrix (the orange elements) leads to a technique called *Principal Components Regression*, or PCR (Hastie et al. 2001). Here, the dimensionality of the **X** space is reduced through PCA, and the resulting components are treated as new predictors for each response in **Y** using MLR. To fit PCR in JMP requires a two-stage process (**Analyze > Multivariate Methods > Principal Components**, followed by **Analyze > Fit Model**). In many instances, though, PLS is a superior choice.

For completeness, we mention that consideration of the 2 x 2 submatrix associated with the **Y**s (the blue elements) along with the 4 x 4 submatrix associated with the **X**s (the orange elements) leads to *Maximum Redundancy Analysis*, MRA (van den Wollenberg 1977). This is a technique that is not as widely used as PLS, PCA, and PCR.

Why Use PLS?

Consistent with the heritage of PLS, let's consider a simulation of a simplified example from spectroscopy. In this situation, samples are measured in two ways: Typically, one is quick, inexpensive, and *online*; the other is slow, expensive, and *offline*, usually involving a skilled technician and some chemistry. The goal is to build a model that predicts well enough so that only the inexpensive method need be used on subsequent samples, acting as a surrogate for the expensive method.

The online measurement consists of a set of intensities measured at multiple wavelengths or frequencies. These measured intensities serve as the values of **X** for the sample at hand. To simplify the discussion, we assume that the technician only measures a single quantity, so that (as in our MLR example in Chapter 2) **Y** is a column vector with the same number of rows as we have samples.

To set up the simulation, run the script SpectralData.jsl by clicking on the correct link in the master journal. This opens a control panel, shown in Figure 4.2.

Figure 4.2: Control Panel for Spectral Data Simulation

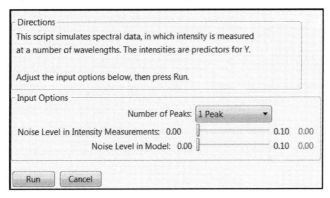

Once you obtain the control panel, complete the following steps:

1. Set the **Number of Peaks** to **3 Peaks**.
2. Set the **Noise Level in Intensity Measurements** slider to 0.02.
3. Leave the **Noise in Model** slider set to 0.00.
4. Click **Run**.

This produces a data table with 45 rows containing an ID column, a Response column, and 81 columns representing wavelengths, which are collected in the column group called Predictors. The data table also has a number of saved scripts.

Let's run the first script, Stack Wavelengths. This script stacks the intensity values so that we can plot the individual spectra. In the data table that the script creates, run the script Individual Spectra. Figure 4.3 shows plots similar to those that you see.

Figure 4.3: Individual Spectra

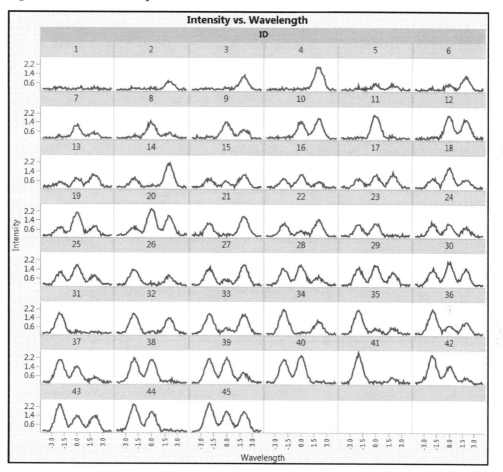

Note that some samples display two peaks and some three. In fact, the very definition of what is or is not a peak can quickly be called into question with real data, and over the years spectroscopists and chemometricians have developed a plethora of techniques to pre-process spectral data in ways that are reflective of the specific technique and instrument used.

Now run the script Combined Spectra in the stacked data table. This script plots the spectra for all 45 samples against a single set of axes (Figure 4.4). You can click on an individual set of spectral readings in the plot to highlight its trace and the corresponding rows in the data table.

Figure 4.4: Combined Spectra

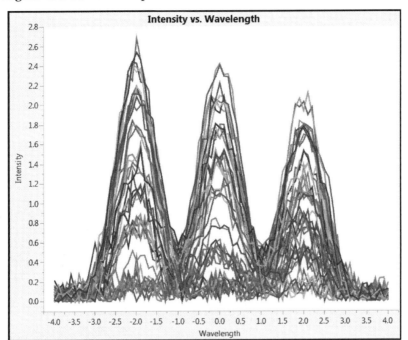

Our simulation captures the essence of the analysis challenge. We have 81 predictors and 45 rows. A common strategy in such situations is to attempt to extract significant features (such as peak heights, widths, and shapes) and to use this smaller set of features for subsequent modeling. However, in this case we have neither the desire nor the background knowledge to attempt this. Rather, we take the point of view that the intensities in the measured spectrum (the row within **X**), taken as a whole, provide a *fingerprint* for that row that we try to relate to the corresponding measured value in **Y**.

Let's close the data table Stacked Data and return to the main data table Response and Intensities. Run the script Correlations Between Xs. This script creates a color map that shows the correlations between every pair of predictors using a blue to red color scheme (Figure 4.5). Note that the color scheme is given by the legend to the right of the plot. To see the numerical values of the correlations, click the red triangle next to **Multivariate** and select **Correlations Multivariate**.

Figure 4.5: Correlations for Predictors Shown in a Color Map

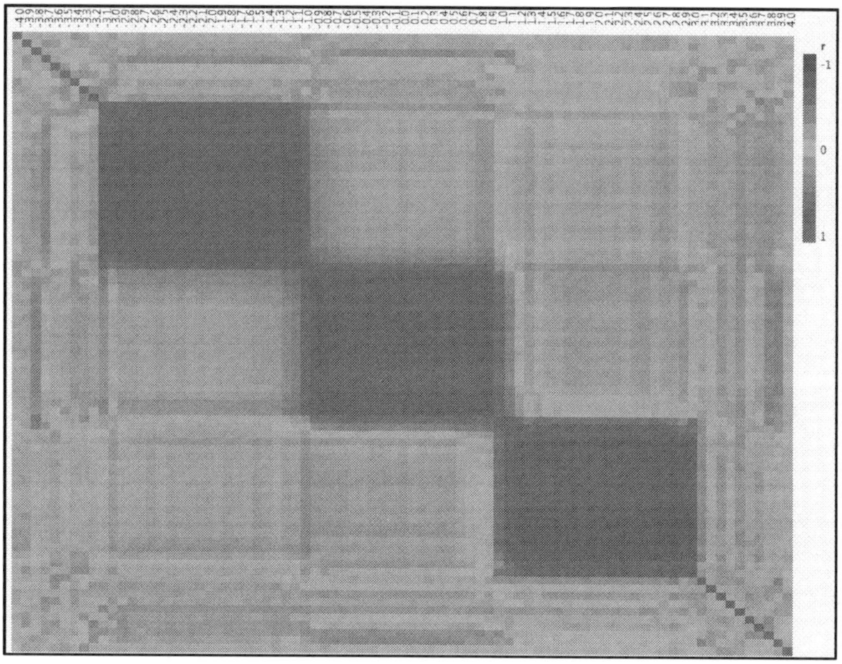

In the section "The Effect of Correlation among Predictors: A Simulation" in Chapter 2, we investigated the impact of varying the correlation between just two predictors. Here we have 81 predictors, one for each wavelength, resulting in 81*80/2 = 3,240 pairs of predictors. Figure 4.5 gives a pictorial representation of the correlations among all 3,240 pairs.

The cells on the main diagonal are colored the most intense shade of red, because the correlation of a variable with itself is +1. However, Figure 4.5 shows three large blocks of red. These are a consequence of the three peaks that you requested in the simulation. You can experiment by rerunning the simulation with a different number of peaks and other slider settings to see the impact on this correlation structure.

Next, in the data table, find and run the script MLR (Fit Model). This attempts to fit a multiple linear regression to **Response**, using all 81 columns as predictors. The report starts out with a long list of **Singularity Details**. This report, for our simulated data, is partially shown in Figure 4.6.

Figure 4.6: Partial List of Singularity Details for Multiple Linear Regression Analysis

```
Singularity Details
Intercept = 17.1265*-4.0 - 14.0426*-3.9 + 4.60419*-3.8 + 16.3727*-3.7 - 15.6405*-3.6 + 19.2797*-3.5 +
21.0897*-3.4 + 8.7617*-3.3 + 0.63614*-3.2 - 3.35463*-3.1 - 12.4783*-3.0 + 3.42323*-2.9 - 3.6529*-2.8 -
4.81473*-2.7 - 2.22472*-2.6 + 12.278*-2.5 + 3.05635*-2.4 - 5.99862*-2.3 + 4.72366*-2.2 - 7.58576*-2.1 +
1.33633*-2.0 - 2.64012*-1.9 - 2.93871*-1.8 + 4.3152*-1.7 - 4.84452*-1.6 + 5.27479*-1.5 + 2.79257*-1.4 -
2.13354*-1.3 + 8.03909*-1.2 - 0.56243*-1.1 + 6.01061*-1.0 + 6.17914*-0.9 - 1.74109*-0.8 + 5.37271*-0.7 +
1.56322*-0.6 - 1.54398*-0.5 + 5.79704*-0.4 - 0.72326*-0.3 + 2.77483*-0.2 + 8.48595*-0.1 - 9.44846*0.0 -
7.26264*0.1 + 14.4864*0.2 + 3.75976*0.3 - 4.77607*0.4 = - 51.7839*-4.0 + 95.3721*-3.9 + 99.0965*-3.8 +
6.61232*-3.7 - 2.05743*-3.6 + 27.3098*-3.5 + 76.6425*-3.4 - 61.6384*-3.3 - 178.615*-3.2 + 84.1121*-3.1 -
62.6616*-3.0 - 74.3761*-2.9 - 25.0727*-2.8 - 53.3957*-2.7 - 42.0132*-2.6 - 9.41151*-2.5 + 51.767*-2.4 -
60.9125*-2.3 + 35.9614*-2.2 + 22.2024*-2.1 - 36.6344*-2.0 - 49.8422*-1.9 - 35.1237*-1.8 + 107.796*-1.7 +
18.7972*-1.6 + 19.233*-1.5 + 39.0155*-1.4 + 30.0125*-1.3 - 8.20854*-1.2 - 0.40636*-1.1 + 70.0569*-1.0 +
40.813*-0.9 - 120.358*-0.8 - 26.8313*-0.7 - 2.71733*-0.6 - 26.0539*-0.5 - 30.2791*-0.4 - 49.8135*-0.3 -
3.09498*-0.2 + 64.8406*-0.1 - 33.2456*0.0 - 48.7058*0.1 + 68.4939*0.2 + 49.7714*0.3 + 34.1862*0.5 =
27.2936*-4.0 - 38.627*-3.9 - 28.9383*-3.8 + 26.3678*-3.7 - 0.17636*-3.6 + 26.3921*-3.5 + 12.2538*-3.4 +
9.5415*-3.3 + 38.6733*-3.2 - 10.5375*-3.1 - 23626*-3.0 + 36.6896*-2.9 + 4.81326*-2.8 + 5.12727*-2.7 +
8.62635*-2.6 + 9.06969*-2.5 - 18.6675*-2.4 - 1.41199*-2.3 - 7.35321*-2.2 - 13.3834*-2.1 + 16.4267*-2.0 +
15.9427*-1.9 + 11.8871*-1.8 - 17.7145*-1.7 + 0.04899*-1.6 - 3.80591*-1.5 - 17.8114*-1.4 - 12.6123*-1.3 +
2.0125*-1.2 + 2.60883*-1.1 - 16.4675*-1.0 + 4.36937*-0.9 + 26.7812*-0.8 + 23.4274*-0.7 - 14.4647*-0.6 +
12.5982*-0.5 + 19.9592*-0.4 + 17.9107*-0.3 + 5.53584*-0.2 - 26.2957*-0.1 + 6.25732*0.0 + 7.55998*0.1 -
16.503*0.2 - 19.1028*0.3 - 17.2762*0.6 = 82.4324*-4.0 - 119.577*-3.9 - 26.8913*-3.8 + 72.2957*-3.7 -
23.5854*-3.6 + 46.0621*-3.5 - 21.2275*-3.4 + 133.682*-3.3 - 100.049*-3.2 - 16.2532*-3.1 + 28.0517*-3.0 +
14.6026*-2.9 + 41.5985*-2.8 - 11.626*-2.7 + 9.88753*-2.6 - 20.8732*-2.5 - 14.0575*-2.4 - 73.2828*-2.3 +
1.35366*-2.2 + 19.4925*-2.1 - 1.1029*-2.0 + 74.6856*-1.9 - 17.6089*-1.8 - 29.6013*-1.7 - 27.7138*-1.6 +
59.63*-1.5 - 63.1425*-1.4 + 52.6333*-1.3 - 26.0566*-1.2 - 5.31276*-1.1 + 84.5677*-1.0 - 18.7129*-0.9 +
5.22534*-0.8 + 75.1704*-0.7 - 17.9287*-0.6 - 33.7181*-0.5 + 4.29591*-0.4 + 66.5545*-0.3 - 34.4704*-0.2 -
51.8839*-0.1 - 5.12079*0.0 - 47.8731*0.1 + 31.6774*0.2 + 29.535*0.3 + 47.0899*0.7 = 14.3843*-4.0 -
7.51273*-3.9 - 12.4131*-3.8 - 7.30604*-3.7 + 5.7987*-3.6 - 5.48889*-3.5 - 9.45039*-3.4 + 17.8085*-3.3 -
2.64832*-3.2 - 5.60417*-3.1 + 13.9826*-3.0 + 7.45266*-2.9 + 12.5433*-2.8 - 2.58096*-2.7 + 1.92259*-2.6 -
11.5167*-2.5 - 12.3886*-2.4 - 8.03067*-2.3 - 4.97041*-2.2 + 3.77772*-2.1 + 0.21592*-2.0 + 16.2915*-1.9 +
6.28059*-1.8 - 8.84459*-1.7 + 4.86679*-1.6 + 1.32132*-1.5 - 7.60588*-1.4 + 6.53807*-1.3 - 2.78414*-1.2 -
5.15439*-1.1 + 1.66653*-1.0 - 2.68456*-0.9 + 7.7244*-0.8 + 12.1932*-0.7 - 4.205*-0.6 - 6.24339*-0.5 +
1.46449*-0.4 + 13.0694*-0.3 - 11.3136*-0.2 - 8.65345*-0.1 + 7.22479*0.0 - 3.44879*0.1 + 3.81116*0.2 -
1.9773*0.3 + 6.21715*0.8 = - 13.8831*-4.0 + 31.1131*-3.9 + 29.1844*-3.8 - 5.54131*-3.7 - 21.838*-3.6 -
10.2014*-3.5 + 26.0768*-3.4 - 18.6296*-3.3 - 2.47391*-3.2 + 0.67042*-3.1 - 10.6136*-3.0 - 28.6804*-2.9 -
14.9481*-2.8 - 10.1923*-2.7 - 7.83288*-2.6 + 13.693*-2.5 + 27.9437*-2.4 + 7.39593*-2.3 + 11.3953*-2.2 -
6.1609*-2.1 - 15.893*-2.0 - 27.9057*-1.9 - 17.5433*-1.8 + 24.551*-1.7 - 0.80982*-1.6 - 4.86271*-1.5 +
36.6396*-1.4 - 1.51827*-1.3 + 14.6935*-1.2 + 5.69932*-1.1 + 15.7846*-1.0 + 9.90834*-0.9 - 29.1241*-0.8 -
18.4094*-0.7 + 6.39545*-0.6 - 0.84587*-0.5 - 6.83586*-0.4 - 28.7386*-0.3 + 12.8199*-0.2 + 12.1542*-0.1 -
```

Here, the **X** matrix has 81+1 = 82 columns, but **X** and **Y** have only 45 rows. Because $n < m$ (using our earlier notation), we should expect MLR to run into trouble. Note that the JMP Fit Model platform does produce some output, though it's not particularly useful in this case. If you want more details about what JMP is doing here, select **Help > Books > Fitting Linear Models** and search for "Singularity Details".

Now run the script Partial Least Squares to see a partial least squares report. We cover the report details later on, but for now, notice that there is no mention of singularities. In fact, the **Variable Importance Plot** (Figure 4.7) assesses the contribution of each of the 81 wavelengths in modeling the response. Because higher *Variable Importance for the Projection* (*VIP*) values suggest higher influence, we conclude that wavelengths between about –3.0 and 1.0 have comparatively higher influence than the rest.

Figure 4.7: PLS Variable Importance Plot

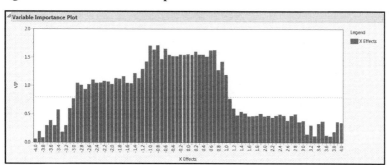

As mentioned earlier, our example is deliberately simplified. It is not uncommon for spectra to be measured at a thousand wavelengths, rather than 81. One challenge for software is to find useful representations, especially graphical representations, to help tame this complexity. Here we have seen that for this type of data, PLS holds the promise of providing results, whereas MLR clearly fails.

You might like to rerun the simulation with different settings to see how these plots and other results change. Once you are finished, you can close the reports produced by the script SpectralData.jsl.

How Does PLS Work?

So what, at a high level at least, is going on behind the scenes in a PLS analysis? We use the script PLSGeometry.jsl to illustrate. This script generates an invisible data table consisting of 20 rows of data with three **X**s and three **Y**s. It then models these data using either one or two factors. Run the script by clicking on the correct link in the master journal. In the control panel window that appears (Figure 4.8), leave the **Number of Factors** set at **One** and click **OK**.

Figure 4.8: Control Panel for PLSGeometryDemo.jsl

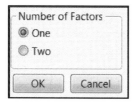

This results in a two-by-two arrangement of 3-D scatterplots, shown in Figure 4.9. A Data Filter window also opens, shown in Figure 4.10.

Because the data table behind these plots contains three responses and $n = 20$ observations, the response matrix **Y** is 20 x 3. This is the first time that we have encountered a matrix **Y** of responses, rather than simply a column vector of responses. To use MLR in this case, we would have to fit a model to each column in **Y** separately. So, any information about how the three responses vary jointly would be lost. Although in certain cases it is desirable to model each response separately, PLS gives us the flexibility to leverage information relative to the joint variation of multiple responses. It makes it easy to model large numbers of responses simultaneously in a single model.

Figure 4.9: 3-D Scatterplots for One Factor

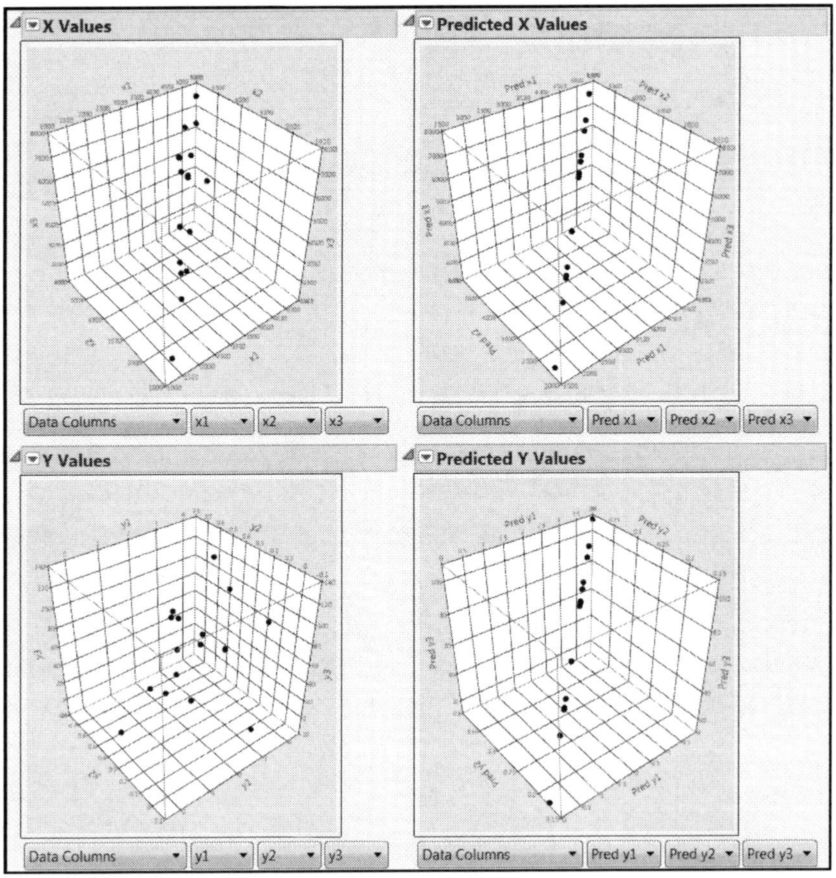

The two 3-D scatterplots on the left enable us to see the actual values for all six variables for all observations simultaneously, with the predictors in the top plot and the responses in the bottom plot. By rotating the plot in the upper left, you can see that the 20 points do not fill the whole cube. Instead, they cluster together, indicating that the three predictors are quite strongly correlated.

You can see how the measured response values relate to the measured predictor values by pressing the **Go** arrow in the **Animation Controls** on the video-like control of the Data Filter window that the script produced (Figure 4.10). When you do this, the highlighting loops over the observations and shows the corresponding observations in the other plots. To pause the animation, click the button with two vertical bars that has replaced the **Go** arrow.

Figure 4.10: Data Filter for Demonstration

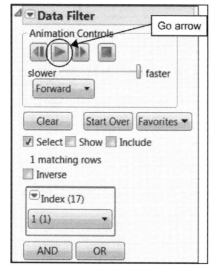

The two plots on the right give us some insight into how PLS works. These plots display predicted, rather than actual, values. By rotating both of these plots, you can easily see that the predicted **X** and **Y** values are perfectly co-linear. In other words, for both the **X**s and the **Y**s, the partial least squares algorithm has projected the three-dimensional cloud of points onto a line.

Now, once again, run through the points using the **Animation Controls** in the Data Filter window and observe the two plots on the right. Note that, as one moves progressively along the line in **Predicted X Values**, one also moves progressively along the line in

Predicted Y Values. This indicates that PLS not only projects the point clouds of the **X**s and the **Y**s onto a lower-dimensional subspace, but it does so in a way that reflects the correlation structure between the **X**s and the **Y**s. If you were given a new observation's three **X** coordinates, PLS would enable you to obtain its predicted X values, and PLS would use related information to compute corresponding predicted Y values for that observation.

In Appendix 1, we give the algorithm used in computing these results. For now, simply note that we have illustrated the statement made earlier in this chapter that PLS is a *projection method* that reduces dimensionality. In our example, we have taken a three-dimensional cloud of points and represented those points using a one-dimensional subspace, namely a line. As the launch window indicates, we say that we have extracted one *factor* from the data. In our example, we have used this one factor to define a linear subspace onto which to project both the **X**s and the **Y**s.

Because it works by extracting latent variables, partial least squares is also called *projection to latent structures*. While the term "partial least squares" stresses the relationship of PLS to other regression methods, the term "projection to latent structures" emphasizes a more fundamental empirical principle: Namely, the underlying structure of highly dimensional data associated with complex phenomena is often largely determined by a smaller number of factors or latent variables that are not directly accessible to observation or measurement (Tabachnick and Fidell 2001). It is this aspect of projection, which is fundamental to PLS, that the image on the front cover is intended to portray.

Note that, if the observations do not have some correlation structure, attempts at reducing their dimensionality are not likely to be fruitful. However, as we have seen in the example from spectroscopy, there are cases where the predictors are necessarily correlated. So PLS actually exploits the situation that poses difficulties for MLR.

Given this as background, close your data filter and 3-D scatterplot report. Then rerun PLSGeometry.jsl, but now choosing **Two** as the **Number of Factors**. The underlying data structure is the same as before. Rotate the plots on the right. Observe that the predicted values for each of the **X**s and the **Y**s fall on a plane, a two-dimensional subspace defined by the two factors. Again, loop through these points using the **Animation Controls** in the Data Filter window. As you might expect, the two-dimensional representation provides a better description of the original data than does the one factor model.

When data are highly multidimensional, a critical decision involves how many factors to extract to provide a sound representation of the original data. This comes back to finding a balance between underfitting and overfitting. We address this later in our examples. For now, close the script PLSGeometry.jsl and its associated reports.

PLS versus PCA

As described earlier, PCA uses the correlation matrix for all variables of interest, whereas PLS uses the submatrix that links responses and predictors. In a situation where there are both **Y**s and **X**s, Figure 4.1 indicates that PCA uses the orange-colored correlations, whereas PLS uses the green-colored correlations. These green entries are the correlations that link the responses and predictors. PLS attempts to identify factors that simultaneously reduce dimensionality and provide predictive power.

To see a geometric representation that contrasts PLS and PCA, run the script PLS_PCA.jsl by clicking on the correct link in the master journal. This script simulates values for two predictors, X1 and X2, and a single response Y. A report generated by this script is shown in Figure 4.11.

Figure 4.11: Plots Contrasting PCA and PLS

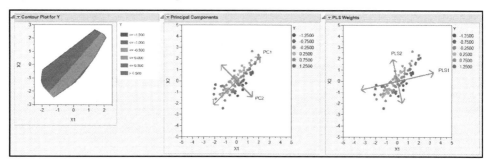

The **Contour Plot for Y** on the left shows how the *true* value of **Y** changes with X1 and X2. The continuous color intensity scale shows large values of **Y** in red and small values in blue, as indicated by the legend to the right of the plot. The contour plot indicates that the response surface is a plane tilted so that it slopes upward in the upper left of the X1, X2 plane and downward in the lower right of the X1, X2 plane. Specifically, the relationship is given by **Y** = –X1 + .75X2.

The next two plots, **Principal Components** and **PLS Weights**, are obtained using simulated values for X1 and X2. But **Y** is computed directly using the relationship shown in the **Contour Plot**.

The **Principal Components** plot shows the two principal components. The direction of the first component, **PC1**, captures as much variation as possible in the values of X1 and X2 regardless of the value of **Y**. In fact, **PC1** is essentially perpendicular to the direction of increase in **Y**, as shown in the contour plot. **PC1** ignores any variation in **Y**. The second component, **PC2**, captures residual variation, again ignoring variation in **Y**.

The **PLS Weights** plot shows the directions of the two PLS factors, or latent variables. Note that **PLS1** is rotated relative to **PC1**. **PLS1** attempts to explain variation in X1 and X2 while also explaining some of the variation in **Y**. You can see that, while **PC1** is oriented in a direction that gives no information about **Y**, **PLS1** is rotated slightly toward the direction of increase (or decrease) for **Y**.

This simulation illustrates the fact that PLS tries to balance the requirements of dimensionality reduction in the **X** space with the need to explain variation in the response. You can close the report produced by the script PLS_PCA.jsl now.

PLS Scores and Loadings

Some Technical Background

Extracting Factors

Before considering some examples that illustrate more of the basic PLS concepts, let's introduce some of the technical background that underpins PLS. As you know by now, a main goal of PLS is to predict one or more responses from a collection of predictors. This is done by extracting linear combinations of the predictors that are variously referred to as latent variables, components, or factors. We use the term *factor* exclusively from now on to be consistent with JMP usage.

We assume that all variables are at least centered. Also keep in mind that there are various versions of PLS algorithms. We mentioned earlier that JMP provides two approaches: NIPALS and SIMPLS. The following discussion describes PLS in general terms, but to be completely precise, one needs to refer to the specific algorithm in use.

With this caveat, let's consider the calculations associated with the first PLS factor. Suppose that **X** is an $n \times m$ matrix whose columns are the m predictors and that **Y** is an

$n \times k$ matrix whose columns are the k responses. The first PLS factor is an $m \times 1$ *weight* vector, w_1, whose elements reflect the covariance between the predictors in X and the responses in Y. The jth entry of w_1 is the weight associated with the jth predictor. The vector w_1 defines a linear combination of the variables in X that, subject to norm restrictions, maximizes covariance relative to all linear combinations of variables in Y. This vector defines the first PLS factor.

The weight vector w_1 is used to weight the observations in X. The n weighted linear combinations of the entries in the columns of X are called *X scores*, denoted by the vector t_1. In other words, the X scores are the entries of the vector $t_1 = Xw_1$. Note that the score vector, t_1, is $n \times 1$; each observation is given an X score on the first factor. Think of the vector w_1 as defining a linear transformation mapping the m predictors to a one-dimensional subspace. With this interpretation, Xw_1 represents the mapping of the data to this one-dimensional subspace.

Technically, t_1 is a linear combination of the variables in X that has maximum covariance with a linear combination of the variables in Y, subject to normalizing constraints. That is, there is a vector c_1 with the property that the covariance between $t_1 = Xw_1$ and $u_1 = Yc_1$ is a maximum. The vector c_1 is a Y *weight* vector, also called a *loading* vector. The elements of the vector u_1 are the *Y scores*. So, for the first factor, we would expect the X scores and the Y scores to be strongly correlated.

To obtain subsequent factors, we use all factors available to that point to predict both X and Y. In the NIPALS algorithm, the process of obtaining a new weight vector and defining new X scores is applied to the residuals from the predictive models for X and Y. (We say that X and Y are *deflated* and the process itself is called *deflation*.) This ensures that subsequent factors are independent of (orthogonal to) all previously extracted factors. In the SIMPLS algorithm, the deflation process is applied to the cross-product matrix. (For complete information, see Appendix 1.)

Models in Terms of X Scores

Suppose that a factors are extracted. Then there are:

- a weight vectors, $w_1, w_2, ..., w_a$
- a X-score vectors, $t_1, t_2, ..., t_a$
- a Y-score vectors, $u_1, u_2, ..., u_a$

We can now define three matrices: W is the $m \times a$ matrix whose columns consist of the weight vectors; **T** and **U** are the $n \times a$ matrices whose columns consist of the X-score and Y-score vectors, respectively. In NIPALS, the Y scores, $\mathbf{u}_i$, are regressed on the X scores, $\mathbf{t}_i$, in an *inner relation* regression fit.

Recall that the matrix **Y** contains k responses, so that **Y** is $n \times k$. Let's also assume that **X** and **Y** are both centered and scaled. For both NIPALS and SIMPLS, predictive models for both **Y** and **X** can be given in terms of a regression on the scores, **T**. Although we won't go into the details at this point, we introduce notation for these predictive models:

(4.1)
$$\hat{\mathbf{X}} = \mathbf{TP}'$$
$$\hat{\mathbf{Y}} = \mathbf{TQ}'$$

where **P** is $m \times a$ and **Q** is $k \times a$. The matrix **P** is called the *X loading* matrix, and its columns are the scaled X loadings. The matrix **Q** is sometimes called the *Y loading* matrix. In NIPALS, its columns are proportional to the Y loading vectors. In SIMPLS, when **Y** contains more than one response, its representation in terms of loading vectors is more complex. Each matrix projects the observations onto the space defined by the factors. (See Appendix 1.) Each column is associated with a specific factor. For example, the ith column of **P** is associated with the ith extracted factor. The jth element of the ith column of **P** reflects the strength of the relationship between the jth predictor and the ith extracted factor. The columns of **Q** are interpreted similarly.

To facilitate the task of determining how much a predictor or response variable contributes to a factor, the loadings are usually scaled so that each loading vector has length one. This makes it easy to compare loadings across factors and across the variables in **X** and **Y**.

Model in Terms of Xs

Let's continue to assume that the variables in the matrices **X** and **Y** are centered and scaled. We can consider the **Y**s to be related directly to the **X**s in terms of a theoretical model as follows:

$$\mathbf{Y} = \mathbf{X}\boldsymbol{\beta} + \boldsymbol{\varepsilon}_Y .$$

Here, $\boldsymbol{\beta}$ is an $m \times k$ matrix of regression coefficients. The estimate of the matrix $\boldsymbol{\beta}$ that is derived using PLS depends on the fitting algorithm. The details of the derivation are given in Appendix 1.

The NIPALS algorithm requires the use of a diagonal matrix, Δ_b, whose diagonal entries are defined by the inner relation mentioned earlier, where the Y scores, $\mathbf{u}_i$, are regressed on the X scores, $\mathbf{t}_i$. The estimate of $\boldsymbol{\beta}$ also involves a matrix, $\mathbf{C}$, that contains the Y weights, also called the Y loadings. The column vectors of $\mathbf{C}$ define linear combinations of the deflated Y variables that have maximum covariance with linear combinations of the deflated X variables.

Using these matrices, in NIPALS, $\boldsymbol{\beta}$ is estimated by

(4.2) $$\mathbf{B} = \mathbf{W}(\mathbf{P'W})^{-1}\Delta_b\mathbf{C'}$$

and Y is estimated in terms of X by

$$\hat{\mathbf{Y}} = \mathbf{XB} = \mathbf{XW}(\mathbf{P'W})^{-1}\Delta_b\mathbf{C'}.$$

The SIMPLS algorithm also requires a matrix of Y weights, also called Y loadings, that is computed in a different fashion than in NIPALS. Nevertheless, we call this matrix $\mathbf{C}$. Then, for SIMPLS, $\boldsymbol{\beta}$ is estimated by

(4.3) $$\mathbf{B} = \mathbf{WC'}$$

and Y is estimated in terms of X by

$$\hat{\mathbf{Y}} = \mathbf{XB} = \mathbf{XWC'}.$$

Properties

Perhaps the most important property, shared by both NIPALS and SIMPLS, is that, subject to norm restrictions, both methods maximize the covariance between the X structure and the Y structure for each factor. The precise sense in which this property holds is one of the features that distinguishes NIPALS and SIMPLS. In NIPALS, the covariance is maximized for components defined on the residual matrices. In contrast, the maximization in SIMPLS applies directly to the centered and scaled X and Y matrices.

The scores, which form the basis for PLS modeling, are constructed from the weights. The weights are the vectors that define linear combinations of the Xs that maximize covariance with the Ys. Maximizing the covariance is directly related to maximizing the correlation. One can show that maximizing the covariance is equivalent to maximizing the product of the squared correlation between the **X** and **Y** structures, and the variance of the **X** structure. (See the section "Bias toward X Directions with High Variance" in Appendix 1, or Hastie et al. 2001.)

Recalling that correlation is a scale-invariant measure of linear relationship, this insight shows that the PLS model is pulled toward directions in **X** space that have high variability. In other words, the PLS model is biased away from directions in the **X** space with low variability. (This is illustrated in the section "PLS versus PCA".) As the number of latent factors increases, the PLS model approaches the standard least squares model.

The vector of X scores, t_i, represents the location of the rows of **X** projected onto the *i*th factor, w_i. The entries of the X loading vector at the *i*th iteration are proportional to the correlations of the centered and scaled predictors with t_i. So, the term *loading* refers to how the predictors relate to a given factor in terms of degree of correlation. Similarly, the entries of the Y loading vector at the *i*th iteration are proportional to the correlations of the centered and scaled responses with t_i. JMP scales all loading vectors to have length one. Note that Y loadings are not of interest unless there are multiple responses. (See "Properties of the NIPALS Algorithm" in Appendix 1.)

It is also worth pointing out that the factors that define the linear surface onto which the **X** values are projected are orthogonal to each other. This has these advantages:

- Because they relate to independent directions, the scores are easy to interpret.
- If we were to fit two models, say, one with only one extracted factor and one with two, the single factor in the first model would be identical to the first factor in the second model. That is, as we add more factors to a PLS model, we do not disturb the ones we already have. This is a useful feature that it is not shared by all projection-based methods; *independent component analysis* (Hastie et al. 2001) is an example of a technique that does not have this feature.

We detail properties associated with both fitting algorithms in Appendix 1.

Example

Now, to gain a deeper understanding of two of the basic elements in PLS, the scores and loadings, open the data table PLSScoresAndLoadings.jmp by clicking on the correct link in the master journal. This table contains two predictors, x1 and x2, and two responses, y1, and y2, as well as other columns that have been saved, as we shall see, as the result of a PLS analysis. The table also contains six scripts, which we run in order.

Run the first script, Scatterplot Matrix, to explore the relationships among the two predictors, x1 and x2, and the two responses, y1, and y2 (Figure 4.12). The scatterplot in the upper left shows that the predictors are strongly correlated, whereas the scatterplot in the lower right shows that the responses are not very highly correlated. (See the yellow cells in Figure 4.12.)

Figure 4.12: Scatterplots for All Four Variables

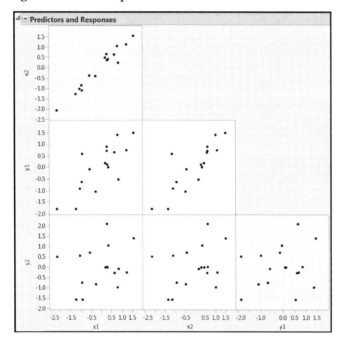

The ranges of values suggest that the variables have already been centered and scaled. To verify this, run the script X and Y are Centered and Scaled. This produces a summary table showing the means and standard deviations of the four variables.

The data table itself includes a number of saved columns that were generated by fitting a PLS model with one factor. If you want to see the details, use the script PLS with One Factor to re-create the report from which the additional columns in the table were saved.

Scores

As we saw in the section "How Does PLS Work?", when we extract a single factor the observations are projected onto a line. The script Projections and Scores helps us explore this projection and the idea of a PLS score. Running the script gives the report shown in Figure 4.13.

Figure 4.13: Plots Describing Projection onto a Line

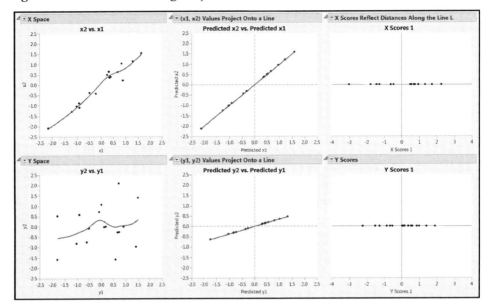

For each of the **X**s and the **Y**s, the leftmost plot shows the raw data, with a smoothing spline superimposed to show the general behavior of the relationship.

The predicted values shown in both middle panels of Figure 4.13 are obtained by regressing the two **X**s and two **Y**s on the score vector **t**, given as X Scores 1 in the data table. Let's focus on the plot for x1 and x2. Because both x1 and x2 are regressed on a single predictor, **t** (X Scores 1), their predicted values fall on a line. Here, we have plotted the predicted values of x1 and x2 against each other. Note that the predicted values do a good job of describing the behavior of the raw data in the **X** space.

Now, let's look at the rightmost plot in the top row. Here we are thinking of the new basis vector defined by the weight vector as having been rotated to a horizontal direction. This axis represents the first PLS factor, and it gives a basis vector for a coordinate system defined by potentially more factors. The points that are plotted along this line are the X scores. Remember, these are the elements of the vector **t** = **Xw**. Note that the X scores reflect precisely the distances from the corresponding points in the middle graph to the origin of the x1 and x2 plane.

Predicted values for y1 and y2 are also obtained by regressing these variables on **t**. The predicted values for y1 and y2 are plotted in the middlemost panel in the second row in Figure 4.13. Again, because both y1 and y2 are regressed on a single predictor, **t**, their predicted values fall on a line. There is more variability in the raw **Y** values than in the raw **X** values, and so the predicted values of y1 and y2 don't represent the data in the **Y** space very closely. Yet, PLS was able to pick up the joint upward trend in the two **Y** variables shown by the smoother.

Recall that the X-score vector, **t**, is determined so as to maximize the covariance with some linear combination of the variables in **Y**. This linear combination of the variables in **Y**, when applied to the observations, produces the vector of Y scores, **u**. These are shown in the rightmost panel of the second row.

To help you see what is happening, while you have the PLS Loadings and Scores window still open, run the script Animation for Scores. This opens a Data Filter window. Run the animation by clicking on the single-headed arrow under **Animation Controls**. Note that you can set the animation speed using the slider in the **Animation Controls** panel.

As the animation runs forward, it scrolls through the rows as they are ordered in the data table. X Scores 1 increases monotonically, but the values of Y Scores 1 do not. This is because, unsurprisingly, the scores themselves are not perfectly related. To see this, run the script X Scores Predicting Y Scores to obtain the plot in Figure 4.14.

Figure 4.14: Y Scores 1 versus X Scores 1

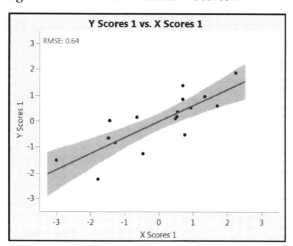

58 *Discovering Partial Least Squares with JMP*

If we had fit a PLS model with *a* latent variables, JMP would allow you to save columns called X Scores 1, X Scores 2, . . ., X Scores a and Y Scores 1, Y Scores 2, …, Y Scores a, back to the data table. As mentioned earlier, we can group the scores $t_1, t_2, ..., t_a$ (respectively, $u_1, u_2, ..., u_a$) into an $n \times a$ matrix **T** (respectively, an $n \times a$ matrix **U**). Note that *scores are associated with rows* in the data table (samples or units on which measurements are made). Scatterplots of scores help diagnose univariate and multivariate outliers, as well as the need to augment the model, perhaps with polynomial terms, to properly account for curvature.

Loadings

Notice the slopes of the blue lines in the two middle plots of Figure 4.13. Given that the horizontal and vertical scales are identical, it appears that x1 and x2 are about equally correlated with the scores on the extracted factor. On the other hand, y1 appears to have a stronger relationship with the scores on the extracted factor than does y2. Confirmatory and more precise information can be found in the report produced by PLS with One Factor. The plots shown in the reports entitled **X Loading Plot** and **Y Loading Plot** are shown in Figure 4.15. These are plots of the X loadings and Y loadings.

Figure 4.15: X and Y Loadings Plots

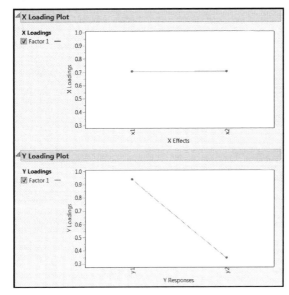

Recall that the Y loadings describe how **Y** is related to the extracted factor, and similarly for the X loadings. Also, as we mentioned earlier, the loadings are normalized so that their values can be compared across factors and across **X** and **Y**. The vertical axes in the two loading plots are identical. You can see that x1 and x2 load about equally on the one extracted factor, while y1 is more important in terms of its loading on that factor than is y2.

Note that if we had fit a PLS model with *a* factors, there would be *a* check boxes (labeled **Factor 1**, **Factor 2**,..., **Factor a**) shown to the left of the plots in this report. This would enable us to view loadings on all factors or only on the checked factors.

Keep in mind that *loadings are related to columns in the data table*, namely to the predictors or responses, and that loadings show the relative importance of each variable in the definition of the extracted factors. Loadings can sometimes help simplify a model by identifying non-essential variables.

An Example Exploring Prediction

To gain more insight on how PLS works as a predictive model, open the data table PLSvsTrueModel.jmp by clicking on the link in the master journal. This table contains two **X**s, three **Y**s, and ten observations. Click on the plus icons next to Y1, Y2, and Y3 in the **Columns** panel to see how the **Y**s are determined by the **X**s. You see the following:

$$Y1 = X1 + RandomNormal(0, 0.01)$$
$$Y2 = X2 + RandomNormal(0, 0.01)$$
$$Y3 = X1 + X2 + RandomNormal(0, 0.01)$$

(The notation *RandomNormal*(0, 0.01) represents normal noise variation with mean 0 and standard deviation 0.01.)

Run the script Multivariate to see the correlation relationships among the five variables (Figure 4.16). The correlations of Y1 with X1 and Y2 with X2 are essentially one, whereas the correlation of Y3 with each of X1 and X2 is almost .99. Beyond this, the **X**s and **Y**s are highly correlated among themselves.

Figure 4.16: Correlations among Xs and Ys

Correlations					
	X1	X2	Y1	Y2	Y3
X1	1.0000	0.9550	1.0000	0.9558	0.9881
X2	0.9550	1.0000	0.9556	0.9999	0.9892
Y1	1.0000	0.9556	1.0000	0.9562	0.9884
Y2	0.9558	0.9999	0.9562	1.0000	0.9896
Y3	0.9881	0.9892	0.9884	0.9896	1.0000

One-Factor NIPALS Model

We begin by fitting a one-factor NIPALS model to the **Y**s. We then obtain the prediction formulas. These are already saved to the data table, but we first show how to obtain them. Then we examine the prediction formulas.

Obtaining the Prediction Formulas

Populate the Fit Model window as illustrated in Figure 4.17 by selecting **Analyze > Fit Model**. Click **Run**.

Figure 4.17: Fit Model Window

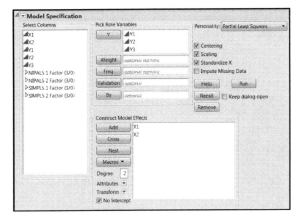

In the Partial Least Squares Model Launch control panel, accept the **NIPALS** default, set the **Validation Method** to **None**, and set the **Initial Number of Factors** to 1 (shown in Figure 4.18). Click **Go**.

Figure 4.18: Partial Least Squares Model Launch Settings

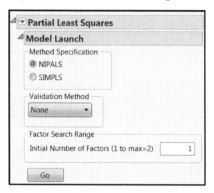

In the **NIPALS Fit with 1 Factors** report, click the red triangle and select **Save Columns > Save Prediction Formula**. This adds three new columns to the far right in your data table. It also appends these three columns to the **Columns** panel at the left of the data grid. Click the + sign to the right of the new column Pred Formula Y1 5 in the **Columns** panel. You should see the formula shown in Figure 4.19. Click the red triangle above the function key pad and select **Simplify**, as shown in Figure 4.19.

Figure 4.19: Formula for Y1

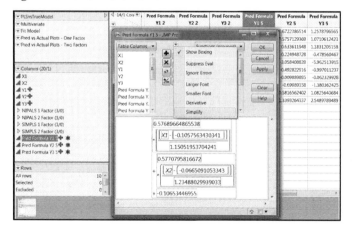

When simplified, with coefficients rounded to four decimal places, the formula is

$$-0.0224 + 0.5014 * X1 + 0.4673 * X2$$

Formulas for the three responses have been obtained in this fashion. They are in the column group called **NIPALS 1 Factor**. Similarly constructed column groups are given for the formulas for the one-factor SIMPLS fit and for the NIPALS and SIMPLS two-factor fits (SIMPLS 1 Factor, NIPALS 2 Factor, and SIMPLS 2 Factor).

Examining the Prediction Formulas

Recall that Y1 is essentially determined by X1, Y2 by X2, and Y3 by the sum of X1 and X2. In the column group **NIPALS 1 Factor**, click on each of the plus signs for the three formulas. Despite the underlying model, the coefficients for the **X**s in all prediction formulas place approximately equal weight on each of X1 and X2. Although the prediction formula for Y3 is consistent with the underlying model, the prediction formulas for Y1 and Y2 are not. Here are the prediction formulas:

$$\text{PredFormula } Y1 = -0.0224 + 0.5014 * X1 + 0.4673 * X2$$
$$\text{PredFormula } Y2 = -0.0225 + 0.5362 * X1 + 0.4997 * X2$$
$$\text{PredFormula } Y3 = -0.0004 + 1.0378 * X1 + 0.9672 * X2$$

Yet, the prediction formulas do a very good job of predicting the **Y** values. Run the script **Pred vs Actual Plots – One Factor**, which produces Figure 4.20. This produces plots of the predicted values against the actual values, together with a line on the diagonal (intercept 0 and slope 1).

Figure 4.20: Predicted by Actual Plots for One-Factor PLS Model

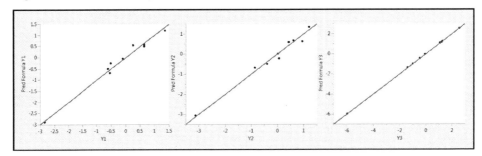

The reason that all three fits are good is that PLS exploits the correlation among the **X**s in determining latent factors. With the single factor it has extracted, PLS models the fairly high correlation between X1 and X2, as well as the covariance with the variables in **Y**. The result is that both X1 and X2 receive about equal weight in the one-factor model.

You can check that standard least squares models for Y1 and Y2 using X1 and X2 as predictors are more faithful to the true model and provide better fits than does PLS.

However, the PLS fit we examined used only one factor (or predictor), while the regression fits use two predictors. If you fit PLS models with two factors, these two factors explain all of the variation in the **X**s and the prediction models for the **Y**s are identical to the standard least squares fits.

This example also illustrates very strikingly how the one-factor PLS model is biased in the direction of maximum variability in the **X**s. Figure 4.21 shows a plot of X2 versus X1. The two variables are highly collinear, and the direction of maximum variability is close to the 45 degree line in the X1 and X2 plane. This is reflected in the coefficients for the one-factor PLS fits, where the coefficients for X1 and X2 are nearly equal, with the coefficient for X2 slightly smaller than the coefficient for X1. As mentioned earlier, when a second factor is added to the PLS model, the prediction models are identical to the standard least squares regression models, and the bias disappears.

Figure 4.21: Scatterplot of X1 versus X2

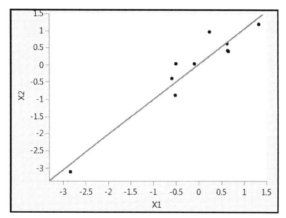

Two-Factor NIPALS Model

Now, we fit a two-factor model. If you run the Fit Model script, you see a report entitled **NIPALS Fit with Two Factors**. We have already saved the prediction formulas for this two-factor fit to the data table and simplified them. Examine these formulas to see that the following prediction equations were obtained:

$$\text{PredFormula } Y1\_2 = -0.0008 + 0.9960 * X1 + 0.0066 * X2$$
$$\text{PredFormula } Y2\_2 = -0.0005 + 0.0099 * X1 + 0.9901 * X2$$
$$\text{PredFormula } Y3\_2 = -0.0007 + 1.0125 * X1 + 0.9907 * X2$$

64 *Discovering Partial Least Squares with JMP*

These formulas faithfully reflect the true relationship between the **Y**s and the **X**s. In fact, these are precisely the predictive models that you would obtain were you to regress each of the **Y**s individually on X1 and X2. The two PLS factors capture all of the variation in the **X**s. Figure 4.22 shows the predicted values plotted against the actual values. (Script is Pred vs Actual Plots – Two Factors.)

Figure 4.22: Predicted by Actual Plots for Two-Factor PLS Model

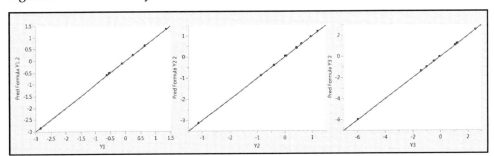

The two-factor PLS model results in better fits than does the one-factor model, but note that the one-factor model predictions were very good, given that only one factor was used. The one-factor model leveraged information about correlation in the **X**s to obtain a very good model for the **Y**s.

Variable Selection

The one-factor model gives us some insight on variable selection using PLS when the modeling objective is geared to explanation rather than prediction. Using the one-factor model, we would be hard-pressed to determine which of X1 or X2 is more likely to be active relative to Y1 or Y2. Because of the high correlation between them, both X1 and X2 have coefficients of about equal size. So, deciding which is active based on coefficient size is not effective. The VIP criterion mentioned in the section "Why Use PLS?" is often used for variable selection. But, in this case, this criterion would also be ineffective because of the high correlation.

SIMPLS Fits

SIMPLS predicted values are given as hidden columns in the data table. If you peruse these, you will not see any differences of note relative to the NIPALS fits. In fact, the one-factor fits are identical—this is always the case. The two-factor prediction equations are identical to the least squares fit.

You can now close **PLSvsTrueModel.jmp**.

Choosing the Number of Factors

Cross Validation

In the section "Underfitting and Overfitting: A Simulation" in Chapter 2, we showed how critical it is to select a model that strikes a balance between fitting the structure in the data and fitting the noise. In terms of PLS, this becomes a question of how many factors to extract. If we were to build a PLS model with the maximum possible number of factors, then we would very likely be overfitting the data.

Our hope is that there is sufficient correlation within the data so that a value of a much smaller than the rank of **X** provides us with a useful model. We have already seen this happen in specific examples. For example, recall that the spearhead data in the section "An Example of a PLS Analysis" in Chapter 1 had 10 predictors. Yet, using a model with $a = 3$, we were able to perfectly classify 10 new spearheads with no errors. In this example, the dimensionality of the **X** space was effectively reduced from 10 to 3. So the general question is, how should we determine the optimal number of factors when conducting a PLS analysis?

Think of a data set as being partitioned into two parts: a *training set* and a *validation set*. The training set is used to develop a model. So, for example, the rows in the training set are used to estimate the model parameters. That model is then evaluated using the validation set. This is done by applying the model to the rows in the validation set and using one or more criteria that measure how well the model predicts this data. In this way, the fit of the model is evaluated on a set of data that is independent of the data used to develop the model.

The value of this strategy is in choosing among models. One constructs a number of different models using the training set. For example, one might compare multiple linear regressions, or partition models, or neural nets, with different predictor sets; or one might compare PLS models with different numbers of factors. These models all tend to fit the training set better than they fit "new" observations, because their parameters are tuned using the training data. To take this bias out of the picture, all of these models should be compared in terms of their performance on the independent validation set. On this basis, a "best" model is chosen.

Although the validation set is independent of the training set that was used to fit the best model, the choice of the best model is nevertheless influenced by the validation set. So, again, there is the potential for bias when one uses that best model to predict values for new observations. To avoid an over-optimistic picture of how this best model will perform on new observations, analysts often reserve a *test set* as well. The test set is used exclusively for assessing the best model's performance. So, in a data-rich situation, one might partition the data, for example, into a 50% training set, a 30% validation set, and a 20% test set.

Think back to the data table Spearheads.jmp that we used as an example in Chapter 1. The model for the spearhead data was built using nine observations, and the performance of this model was assessed using 10 observations that had not been used in the construction of the model. The nine observations were our training and validation set. The remaining 10 observations were treated as a test set, enabling us to assess the usefulness of our model.

Types of Cross Validation

We focus only on training and validation sets at this point. The validation scenario that we have described, using a single training set and a single validation set, is called the *holdout* method. This is because we "hold out" some observations for model validation. The limitation of this method is that the evaluation is highly dependent on exactly which points are included in the training set and which are included in the validation set. One can fall victim to an unfortunate split.

The *k-fold cross validation* method improves on the holdout method. In this method, the data are randomly split into k subsets, or *folds*, of approximately equal size. Each one of the k folds is treated as a holdout sample. In other words, for a given fold, all the remaining data are used as a training set and that fold is used as a validation set. This leads to k analyses and then to k values of the statistic or statistics used to evaluate prediction error. These statistics are usually averaged to give an overall measure of fit.

The k-fold cross validation method ensures that each observation in the data set is used in a validation set exactly once and in a training set $k-1$ times. So how the data are divided is less important than in the holdback method, where, for example, one outlier in the training or validation set can unduly influence conclusions.

A special type of k-fold cross validation consists of choosing k to be equal to the number of observations in the data set, n. This is called *leave-one-out* cross validation. The data are split into n folds, each consisting of a single observation. For each fold, the model is fit to all observations but that one observation, and that single holdout observation is used as the validation set. This leads to n estimates of prediction error that are averaged to give a final measure for evaluation. Note that when all observations but one are used to train the data, the model obtained is unlikely to differ greatly from a model obtained using the entire data set.

The question of which type of cross validation is best for model selection is a difficult one (Arlot and Celisse 2010). It is generally agreed that *k*-fold cross validation is better than the holdout method. In practice, the choice of the number of folds depends on the size of the data set. Often, moderate numbers of folds suffice. The general consensus is that 10 is an adequate number of folds. The leave-one-out method tends to result in error estimates with high variance and is generally used only with small data sets.

A Simulation of K-Fold Cross Validation

Let's get a sense of how *k*-fold cross validation works using the data table BigClassCVDemo.jmp, which you open by clicking on the correct link in the master journal. This is the same data as in the JMP sample data table BigClass.jmp, consisting of measurements on 40 students. We are interested in predicting **weight** from **age**, **sex**, and **height**.

Run the first script, Create Folds. This creates a new column called Folds that randomly assigns each observation to one of four folds in a way that creates folds of equal sizes. Select **Analyze > Distribution** to verify that each fold consists of 10 observations.

Our first step is to fit a regression model to all observations *not* in fold 1. To do this, run the script Fit Model Excluding Fold 1. You see that the script excludes and hides the fold 1 rows, generates a regression report, and adds a column to the data table, Pred Formula weight 1, containing the prediction formula generated by the model. Keep in mind that the prediction formula is developed using a model that was fit, or trained, without the fold 1 observations.

To see how well the model fits observations in the validation set, fold 1, one approach is to consider the *root mean square error of prediction* (RMSE) for fold 1. In other words, you take the differences between the actual values of **weight** and those predicted based on the training set, square these, and take the square root of their adjusted mean. This estimates the error variability. To get a visual sense of this calculation, run the script Graph Builder – Fit Excluding Fold 1. For our random allocation of observations to folds, we obtain the plot in Figure 4.23. Note that your random allocation leads to slightly different results from those shown in Figure 4.23.

Figure 4.23: Comparison of Fits across Folds with Fold 1 as Validation Set

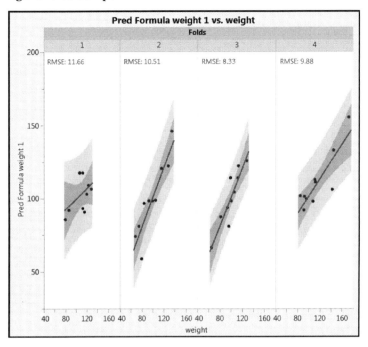

We would expect the fit for fold 1 observations to be the worst, and in this case it is. For our random allocation, fold 1 has the highest **RMSE**. If the data for the other three folds were combined, their combined **RMSE** would be somewhat lower than that of fold 1. All the same, the model fit for fold 1 observations is not that much worse than the fit for the three folds used in the training set.

At this point, we suggest that you run the three remaining Fit Model Excluding Fold "k" scripts to save the prediction formulas to the data table. You can also run the Graph Builder – Fit Excluding Fold "k" scripts. Once you have saved the four prediction formulas, run the script Comparison. This script plots predicted versus actual values for each fold, where the model has been fit using the remaining data. It also plots the actual values of weight and their corresponding predicted values by row. The distances between these points are the prediction errors. These are the values that are squared to form the prediction error sum of squares, used in defining the RMSE.

You see variation in the RMSE over the four folds. The average of these four values provides a measure of how well a multiple linear regression model based on the three predictors fits new observations.

Validation in the PLS Platform

JMP Pro users see all of the validation methods that we have described under **Validation Method** in the PLS Model Launch control panel. Specifically, the selection includes **KFold**, **Holdback**, **Leave-One-Out**, and **None**. (Although **Leave-One-Out** is not in JMP Pro 10, it can be obtained by specifying **KFold** with the number of folds equal to the number of observations.) The default value of k for k-fold cross validation used in the PLS platform is seven, but this can be changed as desired. Users of JMP have a choice of **Leave-One-Out** and **None**.

Assuming that you use one of the JMP validation methods when you run PLS, the PLS platform helps you to identify an optimum number of factors. If you don't use a validation method, you need to specify an **Initial Number of Factors**. Based on the report you obtain, you can then choose to reduce or increase that number of factors. In either case, the maximum number of factors that can be fit is 15, regardless of the number of predictors and responses. Note that this is consistent with the interests of dimensionality reduction.

When you use a model validation method, the PLS platform bases its selection of the optimum number of factors on the *Root Mean PRESS* statistic (where PRESS comes from "**P**redicted **RE**sidual **S**um of **S**quares"), evaluated on validation observations. This is a value that is computed in a fashion similar to how the prediction root mean square error was computed for BigClassCVDemo.jmp earlier. The details of the JMP calculation are given in the section "Determining the Number of Factors" in Appendix 1.

There are a few points to be noted relative to the JMP PLS implementation:

- The PLS platform enables you to fit and review multiple models within the same report window. But, if you use a model validation method, it presents a default fit of a PLS model based on the number of factors that it has determined to be optimum. To fit a model with a different number of factors, you need to specify that number in the Model Launch control panel, which remains open at the top of the report, and run the new analysis.

- Although the word "optimum" is seductive, when your modeling goals involve an aspect of explanation as well as prediction, it might be advantageous to investigate models with a smaller number of factors. Often there are models with fewer factors that are essentially equivalent to the selected optimum model.

- When you use a model validation method, the PLS platform provides you with van der Voet T^2 tests (van der Voet 1994) to help you formally identify models with smaller numbers of factors that are not significantly different from the optimum model. Van der Voet p-values are obtained using Monte Carlo simulation, so the p-value associated with a value of the test statistic is not unique.

The following comments apply specifically to JMP Pro 10.0.2 and 11. With row states, validation columns, and the built-in capabilities of the PLS platform, JMP Pro provides a number of ways to control exactly how rows are used for training, validation, and testing.

- In JMP Pro, PLS can be fit using the **Partial Least Squares** personality in the Fit Model launch window. This window allows for a user-specified **Validation** column to be entered. If you assign a column to this role, it is used to define a holdout sample. If you do not assign a column to this role, then you are presented with **Validation** options in the Model Launch control panel: **KFold**, **Holdback**, **Leave-One-Out** (in JMP Pro 11), and **None**.

- The JMP convention in defining validation columns is as follows: A validation column can contain various integer values. If it contains only two values, say i_1 and i_2, where $i_1 < i_2$, then i_1 corresponds to the training data and i_2 corresponds to the validation data. In JMP Pro 11, if the validation column has three values, $i_1 < i_2 < i_3$, then i_3 corresponds to test set data. If the number of levels, r, exceeds 2 in JMP Pro 10.0.2 or 3 in JMP Pro 11, then the PLS platform uses the r levels to perform r-fold cross validation.

- Note that a validation column with two levels does not result in 2-fold cross validation, because any one row appears in only one group, and the model is only fit once.

- Once a validation strategy has been applied using the Model Launch control panel, you can save a column to the data table containing the fold or holdback assignments for the observations. Select **Save Columns > Save Validation** from the red triangle menu of any one of the fits in the report window. This column can then be used in the **Validation** role in a subsequent Fit Model launch, if desired.

- The random assignments used by **KFold** and **Holdback** are controlled by a random seed. If you need to reproduce the same assignment of observations to validation sets for future PLS launches, you can select **Set Random Seed** from the top level red triangle menu in the Partial Least Squares report, and set the random seed to a specified value.

- If you choose to use a two-level or three-level validation column in the Fit Model window or the validation method **Holdback** in the PLS Model Launch control panel, all models fit in the PLS report window are based on the training data only. If you use **KFold** validation, then the models are fit using all observations.

In the examples in subsequent chapters, we often use row states to isolate a small test set of observations. With the remaining observations, we fit models using JMP validation functionality to build and validate models. A test set that is completely independent of the modeling process is extremely useful for model assessment, and for demonstration and learning purposes.

The NIPALS and SIMPLS Algorithms

The methodology of partial least squares (PLS) was developed in connection with social science modeling in the 1960s with the work of Herman Wold (Dijkstra 2010; Mateos-Aparicio 2011; Wold, H. 1966). Svante Wold, Herman Wold's son, has added substantially to the body of theory and applications connected with PLS, particularly in the area of chemometrics.

The algorithm that Herman Wold implemented was called the *nonlinear iterative partial least squares* algorithm (NIPALS), and this became a foundational element for the theory of PLS regression (Wold, H. 1980). Actually, the NIPALS algorithm existed before it was used for PLS (Wold, H. 1980). The algorithm is based on a sequence of simple ordinary least squares linear regression fits.

By way of review, consider the case where **Y** consists of a single response variable. The first factor is the linear combination of the **X** variables that has the maximum covariance with **Y**. A score vector for the **X** variables is computed by applying this linear combination to the **X** variables, and then **Y** is regressed on that single score vector. The deflated residuals are calculated, and the process is repeated so that a second factor that maximizes the covariance with **Y** is extracted. Working with residuals ensures that the second factor is orthogonal to the first. This process is repeated successively until sufficiently many factors have been extracted.

NIPALS avoids working with the covariance matrix directly. As such, it can handle very large numbers of **X** and **Y** variables. We have seen that cross validation can be used to determine when to stop extracting factors. The idea is that a large amount of the variation in both **Y** and **X** can often be explained by a small number of factors.

The NIPALS algorithm is one method for fitting PLS models. Another popular algorithm is the SIMPLS algorithm, where *SIMPLS* stands for "**S**tatistically **I**nspired **M**odification of the **PLS** Method." This methodology was introduced by Sijmen de Jong in 1993 (de Jong 1993). De Jong's goal was to take a traditional statistical approach to PLS. He wanted to explicitly specify the statistical criterion to be optimized, solve the resulting problem, and then derive an efficient algorithm to generate the solution. He accomplished all of this in his 1993 paper, and the algorithm that he proposed has come to be called the SIMPLS algorithm.

In his paper, de Jong showed that, for a single response variable in **Y**, the NIPALS algorithm and the SIMPLS algorithm lead to identical predictive models. However, for multivariate **Y**, the two predictive models do differ slightly, indicating that the earlier NIPALS method does not exactly optimize de Jong's statistical criterion.

Useful Things to Remember About PLS

As we have seen, PLS is a method for relating inputs, **X**, to outputs, or responses, **Y**, using a multivariate linear model. Its value is in its ability to model data that involve many collinear and noisy variables in both **X** and **Y**, some of which might contain missing data. Unlike ordinary least squares methods, PLS can be used to model data with more **X** and/or **Y** variables than there are observations.

PLS does this by extracting factors from **X** and using these as explanatory variables for **Y**. PLS differs from principal components analysis (PCA), where one reduces the dimension of the **X** matrix by defining factors that are optimal in terms of explaining **X**. In contrast, in PLS, the factors not only explain variation in **X**, but also in **Y**. In PCA, there is no guarantee that a factor will be useful in explaining **Y**. For this reason, many consider PLS a better predictive modeling technique than PCR, which simply regresses each variable in **Y** on the PCA components of **X**.

Partly, but not exclusively, due to the widespread use of automated test equipment, we are seeing data sets with increasingly large numbers of columns (variables, v) and rows (observations, n). Often, it is cheap to increase v and expensive to increase n.

In situations where such a data set consists of predictors and responses, PLS is a flexible approach to building statistical models for prediction. As we have seen, PLS can deal effectively with the following:

- Wide data (when $v \gg n$, and v is large or very large)
- Tall data (when $n \gg v$, and n is large or very large)
- Square data (when $n \sim v$, and n is large or very large)
- Multicollinearity
- Noise

In relative terms, PLS is a sophisticated approach to statistical model building. All the good practice that you already use to build models still applies, but this very sophistication means that you have to be especially diligent to construct a model that is genuinely useful. JMP Pro provides the functionality you need, surfaced in the intuitive style characteristic of JMP. Working through the examples in the next chapters should give you the confidence to use PLS for your own applications.

Predicting Biological Activity

Background 75
The Data 76
  Data Table Description.. 76
  Initial Data Visualization 77
A First PLS Model 79
  Our Plan 79
  Performing the Analysis 79
  The Partial Least Squares Report 81
  The SIMPLS Fit Report 82
  Other Options 83
A Pruned PLS Model 93
  Model Fit 93
  Diagnostics 95
Performance on Data from Second Study 96
  Comparing Predicted Values for the Second Study to Actual Values 96
  Comparing Residuals for Both Studies 99
  Obtaining Additional Insight 101
Conclusion 104

Background

The example in this chapter comes from the field of drug discovery. New drugs are developed from chemicals that are biologically active. Because testing a compound for biological activity is expensive, chemists attempt to predict biological activity from other cheaper chemical measurements. In fact, computational chemistry makes it possible to calculate likely values for certain chemical properties without even making the compound.

In this example, you study the relationship between the size, hydrophobicity, and polarity of key chemical groups at various sites on the molecule, and the activity of the compound. The latter is represented by the logarithm of the relative Bradykinin potentiating activity. We develop a model based on a set of data from one study and then we apply the model to a separate data set from another study. For the first study, you learn that PLS is a useful tool for finding a few underlying factors that account for most of the variation in the response. However, you will also see that the model developed based on the first study's data set does not extend well to the data set from the second study.

The Data

Data Table Description

Open the data table Penta.jmp, partially shown in Figure 5.1, by clicking on the link in the master journal. This table contains 30 rows of observations.

The column obsnam contains an identification code. Each record in Penta.jmp represents a peptide chain of five amino acids. Each amino acid name is coded using a single letter and each chain is represented by five letters, as shown in the column obsnam. The amino acid coding is described in Table 1 of Hellberg et al. (1986).

The response of interest is rai, a relative measure of Bradykinin potentiating activity. (See Table 1 in both Ufkes et al. 1978 and Ufkes et al. 1982). However, rai is highly skewed, and so log_rai, the base 10 logarithm of rai, is used as the response of interest in the analysis. Note that log_rai is given by a formula; click the + sign next to log_rai in the **Columns** panel to view the formula.

The first column in the data table, Study, indicates the study of origin for the given row. The first 15 observations in the table were studied in Ufkes et al. (1978) and the last 15 in Ufkes et al. (1982).

Figure 5.1: Partial View of Penta.jmp

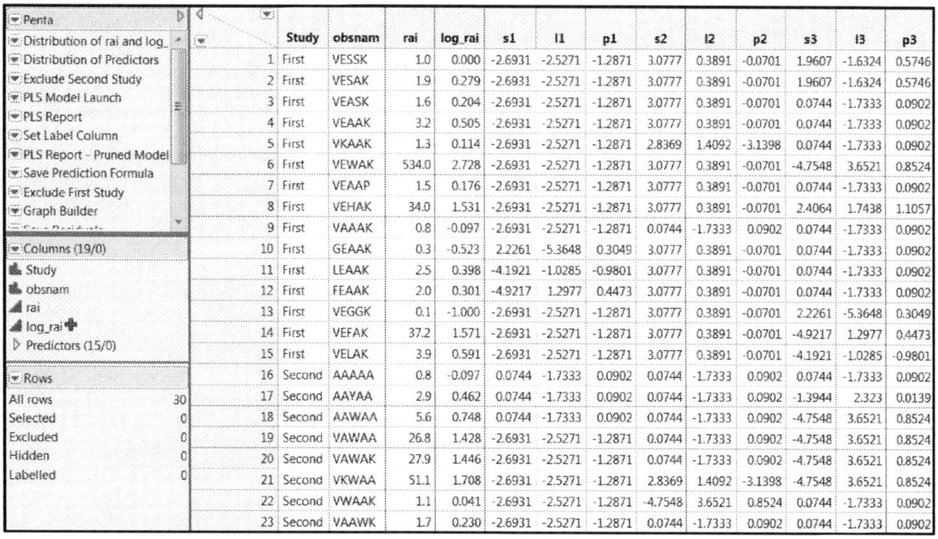

	Study	obsnam	rai	log_rai	s1	l1	p1	s2	l2	p2	s3	l3	p3
1	First	VESSK	1.0	0.000	-2.6931	-2.5271	-1.2871	3.0777	0.3891	-0.0701	1.9607	-1.6324	0.5746
2	First	VESAK	1.9	0.279	-2.6931	-2.5271	-1.2871	3.0777	0.3891	-0.0701	1.9607	-1.6324	0.5746
3	First	VEASK	1.6	0.204	-2.6931	-2.5271	-1.2871	3.0777	0.3891	-0.0701	0.0744	-1.7333	0.0902
4	First	VEAAK	3.2	0.505	-2.6931	-2.5271	-1.2871	3.0777	0.3891	-0.0701	0.0744	-1.7333	0.0902
5	First	VKAAK	1.3	0.114	-2.6931	-2.5271	-1.2871	2.8369	1.4092	-3.1398	0.0744	-1.7333	0.0902
6	First	VEWAK	534.0	2.728	-2.6931	-2.5271	-1.2871	3.0777	0.3891	-0.0701	-4.7548	3.6521	0.8524
7	First	VEAAP	1.5	0.176	-2.6931	-2.5271	-1.2871	3.0777	0.3891	-0.0701	0.0744	-1.7333	0.0902
8	First	VEHAK	34.0	1.531	-2.6931	-2.5271	-1.2871	3.0777	0.3891	-0.0701	2.4064	1.7438	1.1057
9	First	VAAAK	0.8	-0.097	-2.6931	-2.5271	-1.2871	0.0744	-1.7333	0.0902	0.0744	-1.7333	0.0902
10	First	GEAAK	0.3	-0.523	2.2261	-5.3648	0.3049	3.0777	0.3891	-0.0701	0.0744	-1.7333	0.0902
11	First	LEAAK	2.5	0.398	-4.1921	-1.0285	-0.9801	3.0777	0.3891	-0.0701	0.0744	-1.7333	0.0902
12	First	FEAAK	2.0	0.301	-4.9217	1.2977	0.4473	3.0777	0.3891	-0.0701	0.0744	-1.7333	0.0902
13	First	VEGGK	0.1	-1.000	-2.6931	-2.5271	-1.2871	3.0777	0.3891	-0.0701	2.2261	-5.3648	0.3049
14	First	VEFAK	37.2	1.571	-2.6931	-2.5271	-1.2871	3.0777	0.3891	-0.0701	-4.9217	1.2977	0.4473
15	First	VELAK	3.9	0.591	-2.6931	-2.5271	-1.2871	3.0777	0.3891	-0.0701	-4.1921	-1.0285	-0.9801
16	Second	AAAAA	0.8	-0.097	0.0744	-1.7333	0.0902	0.0744	-1.7333	0.0902	0.0744	-1.7333	0.0902
17	Second	AAYAA	2.9	0.462	0.0744	-1.7333	0.0902	0.0744	-1.7333	0.0902	-1.3944	2.323	0.0139
18	Second	AAWAA	5.6	0.748	0.0744	-1.7333	0.0902	0.0744	-1.7333	0.0902	-4.7548	3.6521	0.8524
19	Second	VAWAA	26.8	1.428	-2.6931	-2.5271	-1.2871	0.0744	-1.7333	0.0902	-4.7548	3.6521	0.8524
20	Second	VAWAK	27.9	1.446	-2.6931	-2.5271	-1.2871	0.0744	-1.7333	0.0902	-4.7548	3.6521	0.8524
21	Second	VKWAA	51.1	1.708	-2.6931	-2.5271	-1.2871	2.8369	1.4092	-3.1398	-4.7548	3.6521	0.8524
22	Second	VWAAK	1.1	0.041	-2.6931	-2.5271	-1.2871	-4.7548	3.6521	0.8524	0.0744	-1.7333	0.0902
23	Second	VAAWK	1.7	0.230	-2.6931	-2.5271	-1.2871	0.0744	-1.7333	0.0902	0.0744	-1.7333	0.0902

The data used in this example and a discussion can be found in SAS documentation (*SAS/STAT 9.3 User's Guide*, "The PLS Procedure"). To facilitate comparisons with SAS output, our analysis broadly follows the steps used in the PROC PLS example (Example 69.1). Further background on the data can be found in Ufkes et al. (1978) and Ufkes et al. (1982), Sjostrom and Wold (1985), and Hellberg et al. (1986).

Initial Data Visualization

Let's start by visualizing the data. Run the first saved script, Distribution of rai and log_rai. The plot for rai in Figure 5.2 shows that rai is highly skewed, with some large outlying values.

Figure 5.2: Distribution Reports for rai and log_rai

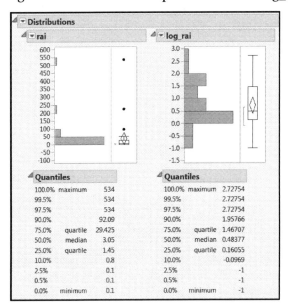

Although PLS does not rely on distributional assumptions of normality, it is still good practice to look at the univariate distributions of the variables and assess whether one of the familiar transformations can make the distribution of a variable assume more of a symmetric, humped shape. In some cases, points that appear to be outliers appear much less extreme on the transformed scale. We see that this is the case here. When rai, the actual response of interest is transformed using a logarithmic function, the distribution of the transformed variable, log_rai, is much more symmetric and well-behaved.

Next, let's visualize the predictors one at a time. You can select **Analyze > Distribution**, or you can run the script called Distribution of Predictors that has been saved to the data table. The plots enable you to get a feel for the data. (See Figure 5.3.) Although some of the distributions appear unruly, given that these are measurements on predictors from a structured study, there is nothing to cause concern.

Figure 5.3: Partial View of Distribution Reports for Penta.jmp Predictors

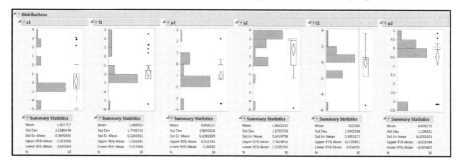

A First PLS Model

Our Plan

We begin by developing a PLS model for the data from the first study, namely the first 15 rows. Recall that these are the amino acid chains that were studied in the Ufkes et al. (1978) paper. We then apply the model we develop to the data from the second study.

To restrict our analysis to only the first 15 rows, we exclude and hide rows 16 to 30. To do this, you can select rows 16 to 30, and then right-click in the highlighted area next to the row numbers and select **Hide and Exclude**. Or, you can run the script Exclude Second Study.

Performing the Analysis

In JMP, PLS is accessed by selecting **Analyze > Multivariate Methods > Partial Least Squares**. It can also be accessed this way in JMP Pro.

1. Enter log_rai as **Y, Response**.
2. Enter all Predictors as **X, Factor**.
3. Click **OK**.

In JMP Pro, PLS can also be accessed through **Fit Model**. Select **Analyze > Fit Model**.

1. Enter log_rai as **Y**.
2. Enter all Predictors as **Model Effects**.
3. Select **Partial Least Squares** as the **Personality**.
4. Deselect the **Standardize X** option.

The **Standardize X** option centers and scales columns that are involved in higher-order terms. Leaving it checked in this case, where we have no higher-order terms, only affects reporting of the model coefficients for the original data. If you access PLS using **Analyze > Multivariate Methods > Partial Least Squares**, you are not able to add higher-order terms, and so the **Standardize X** option is not available.

Your Fit Model window should appear as shown in Figure 5.4.

5. Click **Run**.

Figure 5.4: Fit Model Window

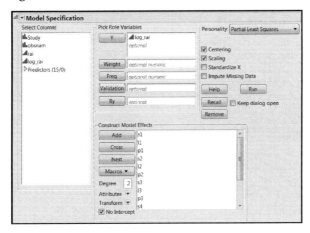

Either of the menu approaches opens the Partial Least Squares Model Launch control panel. In the PLS Model Launch control panel, select **SIMPLS** as the **Method Specification**, choose **None** as the **Validation Method**, and request an **Initial Number of Factors** equal to **2**, as shown in Figure 5.5. You can also run the script PLS Model Launch to obtain the PLS Model Launch control panel.

Figure 5.5: Specification for PLS Model Launch Control Panel

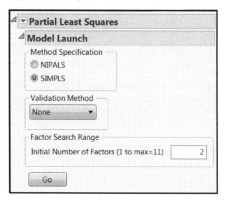

These choices imply that: you are using the SIMPLS algorithm to estimate the model parameters; you are not using model validation within the platform; and you are specifically fitting only two factors to encompass the variation between the model effects (**X**s) and **Y**. Note that we would obtain an identical model had we used NIPALS, because there is only one response. The VIP values, though, would differ slightly.

The Partial Least Squares Report

Clicking **Go** adds additional content to the report. (The script is PLS Report.) Two report sections are appended: A **Model Comparison Summary** report, which is updated when new fits are performed, and a **SIMPLS Fit with 2 Factors** report, showing extensive details about the fit just performed (shown closed in Figure 5.6).

Figure 5.6: PLS Report for Two-Factor Fit

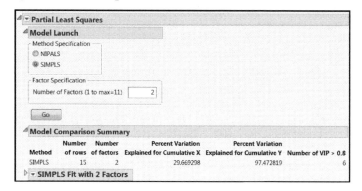

The **Model Comparison Summary** confirms that the model was fit as requested and that the two factors account for about 30% of the variation in the **X**s and 97% of the variation in **Y**. The last column, **Number of VIP > 0.8**, indicates that 6 of the 15 predictors are influential in determining the two factors. This suggests that there might be an opportunity for refining the model by dropping some of the predictors.

Just in passing, we note that the content of JMP reports can be customized using **Preferences**. To customize your PLS options, select **File > Preferences > Platforms > Partial Least Squares Fit** for report options, and **File > Preferences > Platforms > Partial Least Squares** for model launch options. Within the PLS report itself, options can be found in menus obtained by clicking red triangles.

The SIMPLS Fit Report

Now let's look at the report for the fit, **SIMPLS Fit with 2 Factors** (Figure 5.7). The **X-Y Scores Plots** show what we would expect, namely, a correlation between the X and Y scores. In a good PLS model, the first few factors should show a high correlation between the X and Y scores. The correlation usually (but not always) decreases from one factor to the next. The **X-Y Scores Plots** exhibit this behavior nicely.

It would be useful if you could see each observation's amino acid coding as a tooltip when you mouse over a point in the **X-Y Scores Plots**, as illustrated in the plot for the first scores in Figure 5.7. To accomplish this, assign the column obsnam the **Label** attribute. To do this, right-click obsnam in the **Columns** panel and select **Label/Unlabel**. (Alternatively, run the script Set Label Column.)

Incidentally, you can have labels displayed on plots for all or only certain rows by applying the **Label** attribute to the rows of interest. To do this, select the rows in the data table, right-click in the highlighted area, and then select **Label/Unlabel**.

The **Percent Variation Explained** report displays the variation in the **X**s and in **Y** that is explained by each factor. The **Cumulative X** and **Cumulative Y** values must agree with the corresponding figures in the **Model Comparison Summary**.

The two **Model Coefficients** reports give the estimated model coefficients for predicting log_rai.

- The report **Model Coefficients for Centered and Scaled Data** gives the coefficients that apply when the **X**s and **Y** have been centered to have mean zero and standard deviation one.

- The report **Model Coefficients for Original Data** gives coefficients for the model expressed in terms of the raw data values. (Had you checked the **Standardize X** option on the Fit Model launch window, the coefficients in this report would be for a model given in terms of the raw **Y** values but standardized **X** values.)

This second set of model coefficient estimates is often of secondary interest in terms of the analysis.

Figure 5.7: The SIMPLS Fit with 2 Factors Report

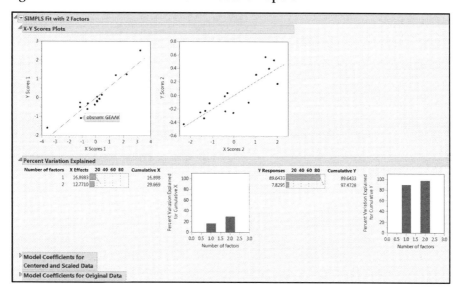

Other Options

A number of useful diagnostic and other tools are available as options from the **SIMPLS Fit with 2 Factors** red triangle menu. We explore a few of these in this section.

Loading Scatterplot Matrices

You can plot the loadings against each other. Recall that, for a given factor, the X loadings reflect the strength of the correlation relationship between the **X** variables and that factor. Similarly, the Y loadings reflect the correlation relationship between the **Y** variables and that factor. Because we have only one **Y** variable, there is a single **Y** loading for each factor. But because we have 15 **X**s, we have 15 loadings for each of the two factors.

Also, keep in mind that loadings are scaled so that, for a given factor, the vector of loadings has length one. This normalization enables us to compare loadings for **X**s and **Y**s across factors.

Select **Loading Scatterplot Matrices** from the red triangle menu for the fit. Figure 5.8 shows the resulting plots.

Figure 5.8: X and Y Loading Scatterplot Matrices

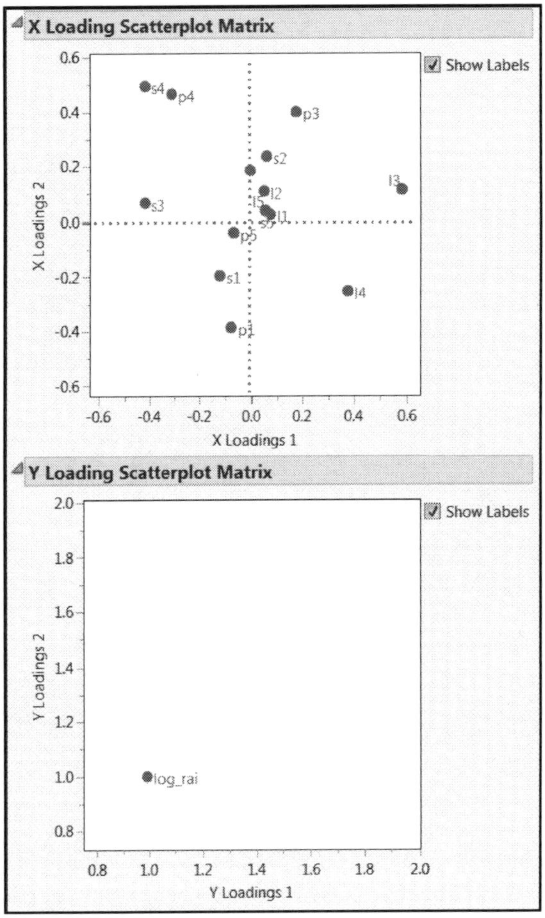

Recall that each column in the projection matrix **P** describes the strength of the correlation between a factor and the predictors. (See Equation (4.1).) In a general situation, the **X Loading Scatterplot Matrix** gives views of these loadings for two factors at a time.

In the **X Loading Scatterplot Matrix** in Figure 5.8, we see that l3 has a high positive loading on the first factor, but a small loading on the second factor. So it is highly correlated with the first factor, but not the second. On the other hand, p1 has a relatively large negative loading on the second factor, but is hardly correlated with the first factor. In fact, the first factor seems to be characterized primarily by the amino acids in the 3rd and 4th positions, while the second factor seems to be characterized by those in the 1st, 3rd, and 4th positions. However, some of these predictors, such as those for the 4th position, have positive correlations with one factor and negative correlations with the other.

This plot also shows a cluster of **X** variables with loadings on both factors that are close to zero. These variables, representing primarily the 2nd and 5th positions, are not explaining much of the variation in the **X** variables.

The **Y Loading Scatterplot Matrix** plots values that represent the scaled correlation between the **Y** variables and each of the two factors. In this case the plot is not particularly informative because there is only one **Y** variable. Because both correlations are scaled to have length one, its loading on both factors is 1.

Loading Plots

Loading plots give another way to view the relationships between the **X**s and **Y**s and the PLS factors. Loading plots are overlay plots that enable you to choose the factors whose loadings you want to display.

Select **Loading Plots** from the red triangle menu for the **SIMPLS Fit with 2 Factors** report. Figure 5.9 shows the two resulting plots. We have added a reference line at 0 to the **X Loading Plot** to help with interpretation. (To do this, double-click on the vertical axis, click **Add** in the **Reference Lines** panel, and then click **OK**.)

The **X Loading Plot** shows that not all predictors impact the factors highly, echoing our conclusions based on the **X Loading Scatterplot Matrix**. In fact, the loadings on both factors for l1, s5, l5, and p5 are close to zero. The predictors s3, l3, s4 and l4, for example, have high loadings, in absolute value, on **Factor 1**, whereas s4, l4, and p4 have high loadings, in absolute value, on both **Factor 1** and **Factor 2**, but in different directions. We think of the two factors as capturing this distinction among these predictors.

Again, because there is only one response variable, the **Y Loading Plot** is uninteresting.

Figure 5.9: Loading Plots

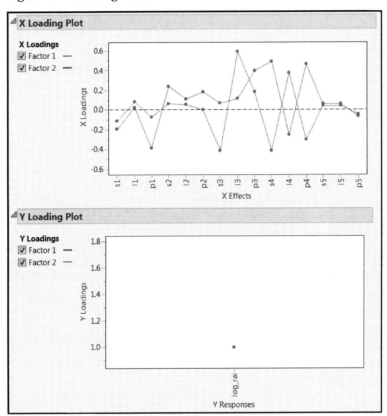

Score Scatterplot Matrices

The X and Y scores express the data in terms of the factors. In this model, there are two factors, so there are two sets of X scores and two sets of Y scores. To look for irregularities relative to the projections of the data to the X and Y spaces, it is useful to plot the X scores and the Y scores against each other. You should look for patterns or clearly grouped observations. If you see a curved pattern, for example, you might want to add a quadratic term to the model. Two or more groupings of observations indicate that it might be better to analyze the groups separately.

To plot scores in this fashion, select **Score Scatterplot Matrices** from the red triangle menu for the **SIMPLS Fit with 2 Factors.** The scatterplot for the X scores is shown in Figure 5.10. The scatterplot matrix has a single cell, because only two factors were extracted. A 95% confidence ellipsoid, calculated using the orthogonality of the X scores, is shown on the plot.

To identify the observations in the scatterplot matrix, select rows 1–15 in the data table, right-click in the highlighted area, and select **Label/Unlabel**. This option tells JMP to label all the selected rows in appropriate plots (Figure 5.10).

Two peptide chains, "VEGGK" and "VEWAK" fall slightly outside the confidence ellipse, although "VEWAK" is essentially on the ellipse boundary. The chain "VEGGK" lies marginally beyond the plotted ellipse, indicating that it might be influential in the PLS analysis. You should check this observation to make sure that it is reliable. Note also that the plot shows some clustering of peptide chains with similar amino acid positional compositions.

Figure 5.10: X Score Scatterplot Matrix

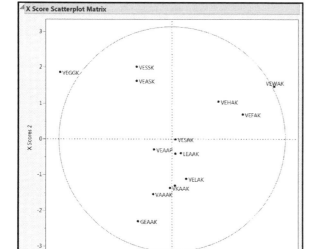

Diagnostics Plots

Selecting **Diagnostics Plots** from the fit's red triangle menu reveals four plots that help in detecting outliers or patterns that might be affecting the fit (Figure 5.11). These plots help you to detect non-normality, autocorrelation, and non-constant variance, all of which can signal problems for the fit.

At this point, we switch to using the row number as a label column. In the **Columns** panel, right-click on obsnam and select **Label/Unlabel** to remove the label attribute from this column. Do not make any changes to the **Label** attribute that is currently applied to rows 1–15. When rows are given the **Label** row state and no column is selected as a **Label** column, the default is to label by row number.

Figure 5.11: Diagnostics Plots

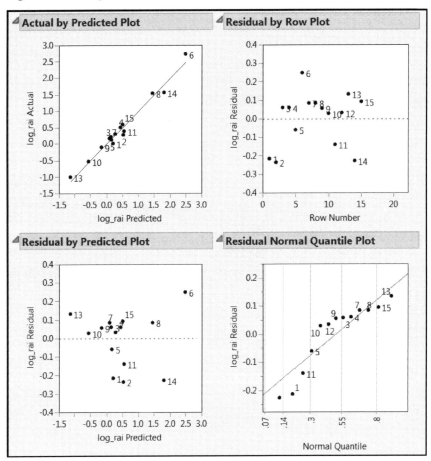

The **Actual by Predicted Plot** shows good agreement between the actual values of log_rai and the values predicted by the two-factor PLS model.

An ideal residual plot looks like a rectangular point cloud with most of the points falling in the vertical middle third of the plot. For these data, neither the **Residual by Predicted Plot** nor the **Residual by Row Plot** shows anything unusual.

In an ideal normal quantile plot, the points fall on a straight line. Here, the **Residual Normal Quantile Plot** shows that several observations are more extreme at the lower end than would be expected. However, this deviation from normality is not serious enough to cause concern.

Variable Reduction in PLS

Recall that PLS models **X** and **Y** by using extracted factors, and then relates **Y** to **X** by fitting a regression model **Y** = **XB** that is derived using those factors. The estimate of the matrix **B** involves the Y loadings as well as the extracted factors. (See Equation (4.2) and Equation (4.3).) It follows that a predictor can be important in the model in at least one of two ways: It can be important in connection with characterizing the factors used to model **X**; or it can be important in terms of the regression model that relates **Y** to **X**. A predictor could be useful in explaining variation in the **X** variables as well as their correlation to **Y**, thus helping to characterize the factors, and yet not be directly useful in predicting **Y**.

The *Variable Importance for the Projection* (VIP) statistic, discussed in Wold (1995, p. 213) and in Wold et al. (2001, p. 123), is defined as a weighted sum of squares of the weights, **W**. (See Appendix 1 and Pérez-Enciso and Tenenhaus 2003.) Being based on the weights, it measures a predictor's contribution to characterizing the factors used in the PLS model, or, equivalently, to defining the projection.

Wold (1995) indicates that predictors that have both small VIP values and regression coefficients near zero can be considered for deletion from the model. Cut-off values for the VIP vary throughout the literature, but there is some agreement that values greater than 1.0 indicate predictors that are important, whereas values below 0.8 indicate predictors that can be deleted, assuming that their regression coefficients are small. In fact, Wold (see Eriksson et al. 2006) suggests a VIP cut-off of 0.8.

Let's summarize our discussion:

- A predictor's VIP represents its importance in determining the PLS projection model for both predictors and responses.
- A predictor's regression coefficient represents that variable's importance in predicting the response.
- If a predictor has a small VIP value and a relatively small regression coefficient (in absolute value), then it is a prime candidate for removal.

There has been considerable study in recent years of variable reduction procedures as they relate to PLS. One should engage in variable deletion cautiously. (See Appendix 2.) Use of cross validation to validate pruned models is prudent.

Variable Importance Plots

We illustrate how variable reduction can be accomplished in our example. We want to look at the VIP for each predictor and at the regression coefficients that make up the **B** matrix (which, in this case, is a column vector because there is only one **Y** and no constant term).

From the red triangle menu for the **SIMPLS Fit with 2 Factors** report, select **Variable Importance Plot**. This option provides both a **Variable Importance Plot** and a **Variable Importance Table** (Figure 5.12). The plot shows the VIP values across the predictors, with a dashed horizontal reference line at 0.8. The **Variable Importance Table** gives a similar plot, but also provides the actual VIP values. You can place these values into a data table by right-clicking in that report, and selecting **Make into Data Table**.

Figure 5.12: Variable Importance Plot and Table

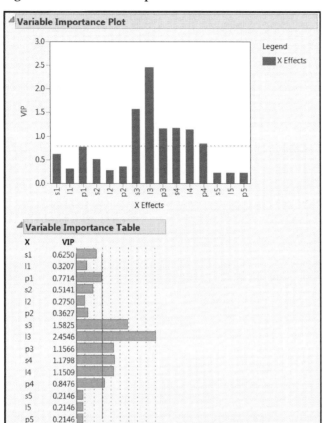

The **Variable Importance Plot** shows six predictors with VIP values exceeding 0.8. One can conclude that these are important for the modeling of both **X** and **Y**. Note that these are measures corresponding to the amino acids in positions 3 and 4 in the peptide chain, which suggests that the amino acids in these positions are important in modeling Bradykinin potentiating energy. The impact of these positions is detailed in Ufkes et al. (1978) and the significance of position 3 is acknowledged in Ufkes et al. (1982).

Now select the option **VIP vs Coefficients Plots** from the red triangle menu for the **SIMPLS Fit with 2 Factors** report. Two plots are shown, one for centered and scaled data and one for the original data. Figure 5.13 shows the plot for centered and scaled data.

The **VIP vs Coefficients** plots show the estimates of the regression coefficients for model terms on the horizontal axis and their VIP values on the vertical axis. Thus, these plots simultaneously give information about how each model term contributes to the regression and to the latent structure. Keep in mind that the values of the regression coefficients are affected by the centering and scaling of the measurements. So, unless there is a compelling reason to do otherwise, we recommend that the user focus on the plot in the **VIP vs Coefficients for Centered and Scaled Data** report, rather than the plot in the **VIP vs Coefficients for Original Data** report, in making decisions about variable reduction.

Figure 5.13: VIP versus Coefficients Plot for Centered and Scaled Data

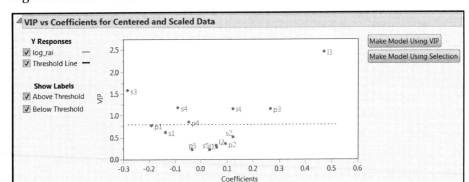

To the right of the plot entitled **VIP vs Coefficients for Centered and Scaled Data**, you see two selection buttons: **Make Model Using VIP** and **Make Model Using Selection**. These buttons offer convenient ways to specify reduced models. If you have fit your model using the Partial Least Squares platform under the Multivariate Methods menu, clicking **Make Model Using VIP** opens a Partial Least Squares launch window where: the **X, Factor** list is populated with the effects whose VIPs exceed 0.8; the **Y, Response** list includes the previously selected responses. If you have fit your model using the Partial Least Squares personality under Fit Model, clicking **Make Model Using VIP** opens an appropriately populated Fit Model launch window.

Alternatively, you might prefer to select predictors directly in the plot by dragging a rectangle or by using the *Lasso* tool. Then, clicking on **Make Model Using Selection** opens a Partial Least Squares launch window where the **X, Factor** list is populated with these selected effects, or a Fit Model launch window where the **Construct Model Effects** list is populated with the selected effects.

We see that the predictors l1, s2, l2, p2, s5, l5, and p5 have small absolute coefficients and small VIPs. Looking back at the **Loading Scatterplot Matrices**, you see that these variables have loadings near zero for both PLS components, indicating that they don't have much influence on the factors that were used to construct the model.

A Pruned PLS Model

Model Fit

Based on our study of VIPs and regression coefficients, we remove the variables l1, s2, l2, p2, s5, l5, and p5 from the PLS model. Note that we do not remove s1 and p1, although their VIPs are below 0.8. These two variables have regression coefficients that are perhaps not negligible, and we prefer to err in the direction of not removing potentially active predictors. So we refit the model with the remaining eight predictors: s1, p1, s3, p3, l3, s4, l4, and p4.

1. In the plot entitled **VIP vs Coefficients for Centered and Scaled Data**, drag a rectangle starting at the upper right of the plot, above and to the right of l3, to include all eight of these variables (s1, p1, s3, p3, l3, s4, l4, and p4), as shown in Figure 5.14. Note that we have resized the plot to make it easier to select the desired variables by dragging a rectangle.

2. Click **Make Model Using Selection**. In the launch window that appears, make sure that you selected the correct variables. If you accidentally missed a variable or added an undesired variable, you can make an adjustment in this window.

3. Click **Run** or **OK**.

Alternatively, you can add the eight variables directly by selecting **Analyze > Multivariate Methods > Partial Least Squares** in both JMP and JMP Pro, and **Analyze > Fit Model** in JMP Pro.

Figure 5.14: Selection of Eight Variables for Pruned Model

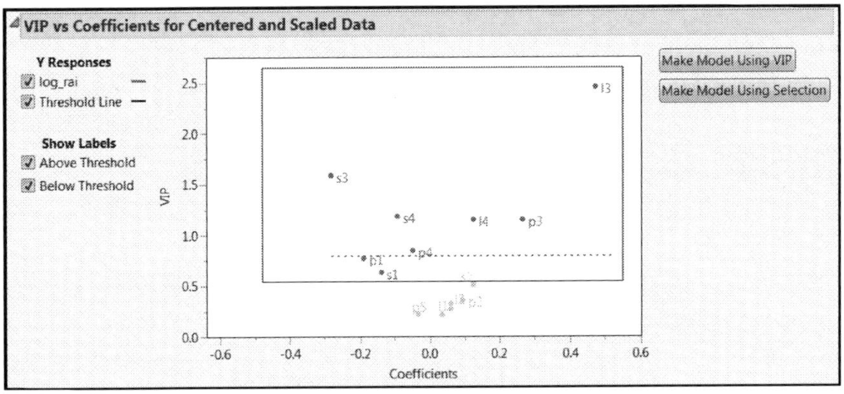

In the PLS Model Launch window, as before, select **SIMPLS** as the **Method Specification**, choose **None** as the **Validation Method**, and request an **Initial Number of Factors** equal to **2**. These choices produce the report shown in Figure 5.15. You can also simply run the script PLS Report – Pruned Model to obtain the report shown in Figure 5.15.

Figure 5.15: SIMPLS Fit with 2 Factors Report for Reduced Model

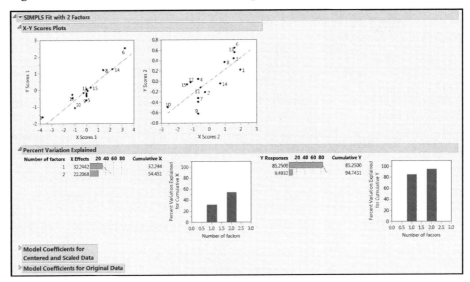

The variation explained in **Y** by the pruned model is about 95%, a slight drop from 97% in the model with all predictors. However, the variation explained in **X** by the pruned model is about 54%, compared with about 30% in the full model. It appears that, by dropping predictors that are not highly related to **Y**, the new PLS factors provide a better representation of the variability in the reduced **X** space.

Diagnostics

From the red triangle menu next to the model fit, select **Diagnostics Plots**. The **Actual by Predicted Plot** shows that the model is predicting well. The **Residual by Predicted Plot** shows no anomalies or patterns. We note that the **Residual Normal Quantile Plot** is closer to being linear than it was for the full model (Figure 5.16).

Figure 5.16: Diagnostics Plots for Pruned Model

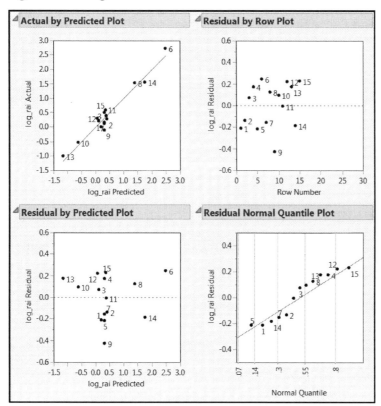

Another way to check for outliers in the model is to look at the Euclidean distance from each observation to the PLS model in both the **X** and **Y** spaces. No observation should be dramatically farther from the model than the rest. Such behavior might indicate that the point is unduly influencing the fit of the model. If there is a group of points that are all farther from the model than the rest, it might be that they have something in common and should be analyzed separately.

Select **Distance Plots** from the red triangle menu to obtain the plots shown in Figure 5.17. With the possible exception of row 9 (which could be further investigated), there appear to be no outliers. Note that these distances to the model are called *DModX* and *DModY* by Umetrics and others (Eriksson et al. 2006).

Figure 5.17: Distance Plots for Pruned Model

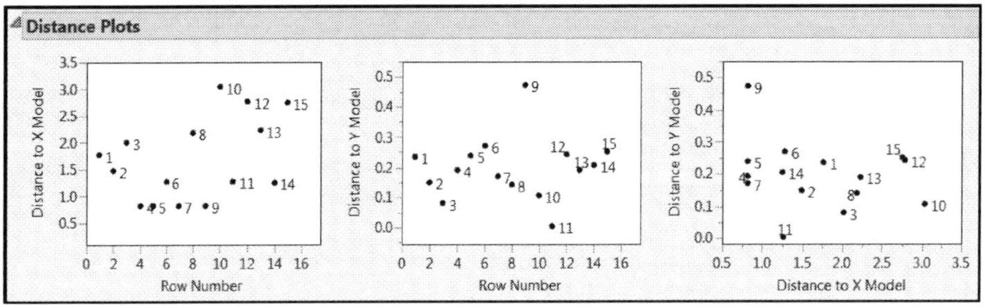

Performance on Data from Second Study

Comparing Predicted Values for the Second Study to Actual Values

The reduced model appears to be more satisfactory than the model including all predictors. So let's see how it performs on the observations that are currently hidden and excluded.

Select **Save Columns > Save Prediction Formula** from the red triangle menu for the **SIMPLS Fit with 2 Factors** report. (The script is Save Prediction Formula.) This option adds a new formula column called Pred Formula log_rai to the data table. Note that predicted values for all 30 observations appear in the data table, because a formula is being saved to the column.

We want to evaluate performance on rows 16–30, which contain the data from the later 1982 study. So, select rows 16–30 in the data table. Right-click on one of the highlighted rows and select **Clear Row States** from the menu that appears. Next, select rows 1–15 in the data table. Right-click in the highlighted area and select **Hide and Exclude**. (Alternatively, run the script Exclude First Study.) Your data table should appear as shown in Figure 5.18.

Figure 5.18: Row States That Define the Second Study

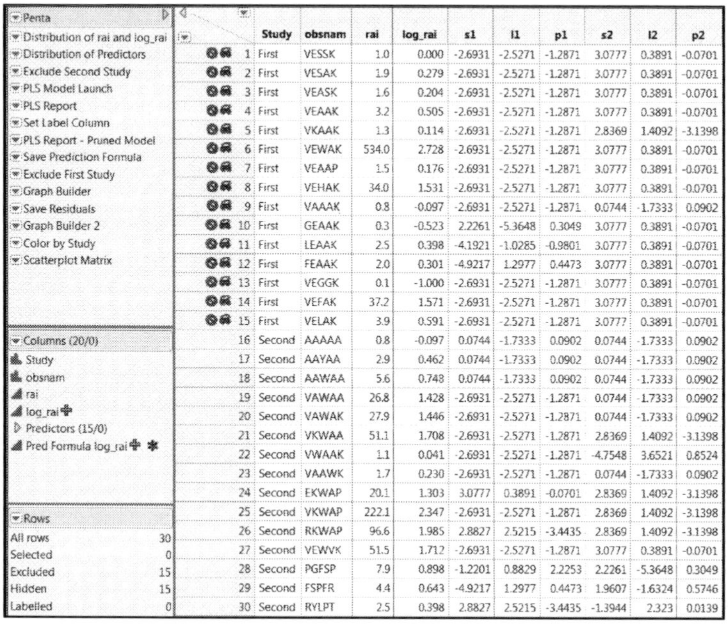

	Study	obsnam	rai	log_rai	s1	l1	p1	s2	l2	p2
1	First	VESSK	1.0	0.000	-2.6931	-2.5271	-1.2871	3.0777	0.3891	-0.0701
2	First	VESAK	1.9	0.279	-2.6931	-2.5271	-1.2871	3.0777	0.3891	-0.0701
3	First	VEASK	1.6	0.204	-2.6931	-2.5271	-1.2871	3.0777	0.3891	-0.0701
4	First	VEAAK	3.2	0.505	-2.6931	-2.5271	-1.2871	3.0777	0.3891	-0.0701
5	First	VKAAK	1.3	0.114	-2.6931	-2.5271	-1.2871	2.8369	1.4092	-3.1398
6	First	VEWAK	534.0	2.728	-2.6931	-2.5271	-1.2871	3.0777	0.3891	-0.0701
7	First	VEAAP	1.5	0.176	-2.6931	-2.5271	-1.2871	3.0777	0.3891	-0.0701
8	First	VEHAK	34.0	1.531	-2.6931	-2.5271	-1.2871	3.0777	0.3891	-0.0701
9	First	VAAAK	0.8	-0.097	-2.6931	-2.5271	-1.2871	0.0744	-1.7333	0.0902
10	First	GEAAK	0.3	-0.523	2.2261	-5.3648	0.3049	3.0777	0.3891	-0.0701
11	First	LEAAK	2.5	0.398	-4.1921	-1.0285	-0.9801	3.0777	0.3891	-0.0701
12	First	FEAAK	2.0	0.301	-4.9217	1.2977	0.4473	3.0777	0.3891	-0.0701
13	First	VEGGK	0.1	-1.000	-2.6931	-2.5271	-1.2871	3.0777	0.3891	-0.0701
14	First	VEFAK	37.2	1.571	-2.6931	-2.5271	-1.2871	3.0777	0.3891	-0.0701
15	First	VELAK	3.9	0.591	-2.6931	-2.5271	-1.2871	3.0777	0.3891	-0.0701
16	Second	AAAAA	0.8	-0.097	0.0744	-1.7333	0.0902	0.0744	-1.7333	0.0902
17	Second	AAYAA	2.9	0.462	0.0744	-1.7333	0.0902	0.0744	-1.7333	0.0902
18	Second	AAWAA	5.6	0.748	0.0744	-1.7333	0.0902	0.0744	-1.7333	0.0902
19	Second	VAWAA	26.8	1.428	-2.6931	-2.5271	-1.2871	0.0744	-1.7333	0.0902
20	Second	VAWAK	27.9	1.446	-2.6931	-2.5271	-1.2871	0.0744	-1.7333	0.0902
21	Second	VKWAA	51.1	1.708	-2.6931	-2.5271	-1.2871	2.8369	1.4092	-3.1398
22	Second	VWAAK	1.1	0.041	-2.6931	-2.5271	-1.2871	-4.7548	3.6521	0.8524
23	Second	VAAWK	1.7	0.230	-2.6931	-2.5271	-1.2871	0.0744	-1.7333	0.0902
24	Second	EKWAP	20.1	1.303	3.0777	0.3891	-0.0701	2.8369	1.4092	-3.1398
25	Second	VKWAP	222.1	2.347	-2.6931	-2.5271	-1.2871	2.8369	1.4092	-3.1398
26	Second	RKWAP	96.6	1.985	2.8827	2.5215	-3.4435	2.8369	1.4092	-3.1398
27	Second	VEWVK	51.5	1.712	-2.6931	-2.5271	-1.2871	3.0777	0.3891	-0.0701
28	Second	PGFSP	7.9	0.898	-1.2201	0.8829	2.2253	2.2261	-5.3648	0.3049
29	Second	FSPFR	4.4	0.643	-4.9217	1.2977	0.4473	1.9607	-1.6324	0.5746
30	Second	RYLPT	2.5	0.398	2.8827	2.5215	-3.4435	-1.3944	2.323	0.0139

To visually compare log_rai and Pred Formula log_rai:

1. Select **Graph > Graph Builder**.
2. Drag log_rai from the **Variables** list to the area under the graph template indicated by **X**.
3. Drag Pred Formula log_rai to the **Y** position next to the vertical axis at the left of the graph template.
4. Click the **Line of Fit** icon above the graph template (third icon from the left).
5. From the **Line of Fit** panel at the lower left, deselect **Confidence of Fit**, as it is not relevant in this case.

This produces the graph shown in Figure 5.19.

Figure 5.19: Predicted versus Actual Values of log_rai for Pruned Model

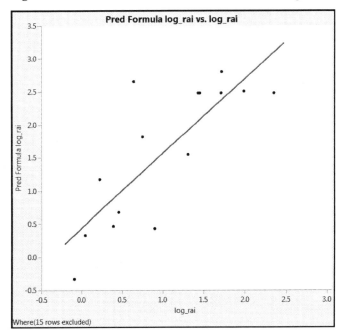

If **Pred Formula log_rai** were predicting log_rai exactly, the actual and predicted values would fall on a diagonal line. But, in fact, the line along which the points fall has slope greater than one.

For a better view, run the script **Graph Builder**. The resulting plot (Figure 5.20) shows a green line plotted at the diagonal. If the model were to fit the data well, the points should fall along this line. On average, the predicted values are higher than the actual values, some by as much as two log_rai units. Although the ultimate value of the model depends on whether differences of this magnitude are important, it appears that the model that was developed using data from the first study shows bias relative to predicting the data from the second study.

Figure 5.20: Predicted versus Actual Values with Line at Diagonal

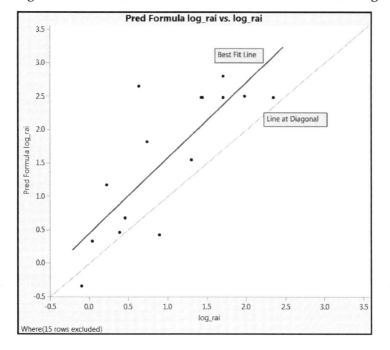

Comparing Residuals for Both Studies

We can gain additional insight by comparing the predicted values to the actual values for both sets of data. There are many ways to do this. We construct a residual plot showing both sets of observations.

To make a new formula column to calculate the residuals for all the data, complete the following steps. (The script is Save Residuals.)

1. To the right of the existing columns in the data table, double-click in the header area to add a new column.
2. Right-click in the header area and select **Formula**.
3. Enter the formula shown in Figure 5.21. Select log_rai in the **Table Columns** list, click the – sign on the operator pad to the right of the column list, and then select Pred Formula log_rai from the **Table Columns** list.
4. Click **OK**.
5. Click on the header for the new column and name it Residuals.

Figure 5.21: Formula for Residuals Column

Now let's use **Graph Builder** to construct the plot shown in Figure 5.22. (The script is Graph Builder 2.)

1. Select **Rows > Clear Row States** to remove the excluded and hidden states.
2. Select **Graph > Graph Builder**.
3. Drag Pred Formula log_rai from the **Variables** list to the area under the graph template indicated by **X**.
4. Drag Residuals to the **Y** position next to the vertical axis at the left of the graph template.
5. Drag Study to the **Color** box to the right of the graph template.
6. Deselect the **Smoother** icon above the graph template (second icon from the left).
7. Double-click on the vertical axis to open the **Y Axis Specification** menu. In the **Reference Lines** panel at the bottom, click **Add** to add a reference line at 0. Click **OK**.
8. If you want, drag the vertical axis settings to center the line at 0.
9. If you want to make the markers larger, right-click in the graph, select **Graph > Marker Size**, and then select the desired size. (We have selected **3, Large**.)
10. From the red triangle menu, deselect **Show Control Panel** to remove the control panel.

This produces the graph shown in Figure 5.22. It is clear that the model fits the data from the first study better than the data from the second study. Again we see that, for the second-study data, the predicted values tend to be too high, especially for larger predicted values.

Figure 5.22: Residual Plot for First and Second Study Data

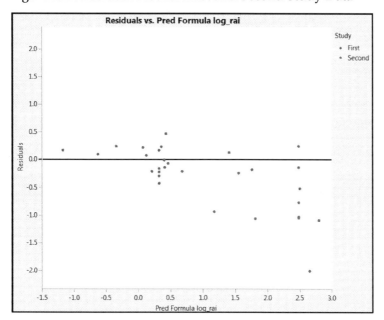

Obtaining Additional Insight

This observation leads us to suspect that the two groups of observations are systematically different in some sense. To gain some insight, we construct a scatterplot matrix. But first, let's color the rows so that we can differentiate between data from the two studies in our scatterplot matrix. (The script is Color by Study, but note that the script does *not* create a legend window.)

1. Select **Rows > Color or Mark by Column**.
2. In the window that opens, select the column Study.
3. In that same window, check **Make Window with Legend**.
4. Click **OK**.

A small portable legend window appears, showing that the color red is associated with the training set and blue with the test set.

The following steps explain how to construct our scatterplot matrix in JMP 11. (The script is Scatterplot Matrix.) The steps in JMP 10 differ slightly.

1. Select **Graph > Scatterplot Matrix**.
2. From the Predictors column group, select s1, p1, s3, l3, p3, s4, l4, and p4, and enter these as **Y, Columns**.
3. Click **OK**.
4. From the red triangle menu in the resulting report, select **Group By**. Click **Grouped by Column** and select Study. These choices cause subsequent analyses to be run separately for each specified group.
5. Click **OK**. This brings you back to the **Scatterplot Matrix** report.
6. Select **Density Ellipses > Density Ellipses** and **Density Ellipses > Shaded Ellipses** from the red triangle menu options.

The scatterplot matrix is shown in Figure 5.23. You can click on the study names in the legend window to highlight the corresponding points in the scatterplot matrix. Note that the observations from the test set (second study) have measurements on the amino acid properties that vary more extensively than the observations for the training set (first study). This plot confirms our suspicions that the two sets of observations are different in some intrinsic way.

Figure 5.23: Scatterplot Matrix for First and Second Study Data

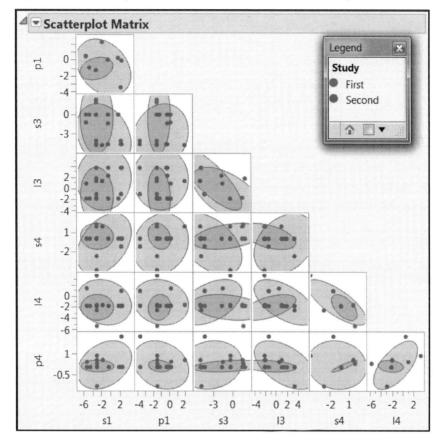

Now recall that the first 15 observations were from the study reported in Ufkes et al. (1978), while the last 15 were from the study reported in Ufkes et al. (1982). It appears that the second study used a broader range of amino acids than did the first. In addition, the second study integrated more variation in the positioning of amino acids in the peptide chain than did the first study. Were we to use our training set model to make predictions relative to the test set, we would be guilty of extrapolation.

In addition to noting differences in the peptides used in the study design, the authors indicate that the Bradykinin used in the two studies came from different sources. This fact might have had an impact on the response measure. Also, when studies conducted at two distinct time points are compared, so-called *lurking variables* (differences in setup, control variables, or measurement procedures) can come into play. So the fact that the model based on the training data does not generalize well to the test data in this case is perhaps not surprising.

At this point, we encourage you to derive a model using all 30 observations. You might want to use cross validation to determine the optimal number of factors. Do you obtain a good predictive model? Note that, with more factors, the differences in the two studies are more adequately modeled.

Conclusion

We have seen how to fit a PLS model to the subset of the data set Penta.jmp reflecting the 1978 study. We used knowledge about regression coefficients, VIPs, and loadings to prune that model. We then applied that model to the data from the 1982 study and found that it did not have good predictive ability. We concluded that the lack of predictive ability might be due to the comparatively smaller predictor range of observations from the first study, or to the impact of lurking variables, or perhaps to some other fundamental difference.

These findings raise two important points relative to modeling:

- In any empirical model building, interpolation is reasonable, but extrapolation is never reasonable.
- The nature and quality of the data is of paramount importance to building sound models. Good models require valid and representative data.

We also note that the dynamic visualization capabilities of JMP are key to developing the insights that help you make sound modeling decisions.

Predicting the Octane Rating of Gasoline

Background 106
The Data 106
    Data Table Description 106
    Creating a Test Set Indicator Column 107
Viewing the Data 108
    Octane and the Test Set 108
    Creating a Stacked Data Table 109
    Constructing Plots of the Individual Spectra 111
    Individual Spectra 112
    Combined Spectra 113
A First PLS Model 116
    Excluding the Test Set 116
    Fitting the Model 117
    The Initial Report 118
A Second PLS Model 120
    Fitting the Model 120
    High-Level Overview 120
    Diagnostics 121
    Score Scatterplot Matrices 125
    Loading Plots 127
    VIPS 129
    Viewing VIPs and Regression Coefficients for Spectral Data 131
    Model Assessment Using Test Set 133
A Pruned Model 136

Background

The example presented in this section deals with predicting the octane rating of gasoline samples based on spectral analysis. The data set contains octane ratings (**Y**) and 401 diffuse reflectance measurements (**X**s), measured over the wavelength range of 900 to 1700 nm in 2 nm increments (that is, in the near infrared range of the spectrum). The goal is to use the spectral data to predict octane content. The data are discussed in Kalivas (1977).

These data are typical spectroscopic data, characterized by many potential predictors (401) measured on a comparatively small number of observations (60). In fact, there has been a large research effort in chemometrics aimed at building calibration models using spectral data and other molecular descriptors. "Fat matrices," matrices where there are many more predictors than observations, characterize these situations. As we saw in the section "Why Use PLS" in Chapter 4, MLR is not capable of fitting a unique model when the design matrix has more predictors than observations.

A least squares MLR fit would require variable selection, and this would likely underutilize the available information. Even then, multicollinearity would cause the least squares fit to be unstable with high prediction variance. This typifies the two main issues associated with fat matrices, namely, rank and multicollinearity.

PLS has two major advantages:

- It is efficient in utilizing the information in the data.
- It effectively *shrinks* the estimates in the coefficient vector or matrix away from the least squares estimates by not including insignificant factors in the model.

And because the extracted factors are mutually orthogonal, the regression that is performed as part of PLS is guaranteed to be non-problematic.

The Data

Data Table Description

Open the Gasoline.jmp data table by clicking on the correct link in the master journal. A portion of the data table is shown in Figure 6.1. Each row gives data on a gasoline sample, whose Sample Number is given in the first column. Octane ratings are in the second column. In the **Columns** panel, the 401 near infrared reflectance (NIR) measurements are in a column group called NIR Wavelengths. To be precise, the values in the NIR columns are transformed values of reflectance, R, given as log(1/R).

Figure 6.1 Portion of Gasoline.jmp Data Table

Sample Number	Octane	NIR.900 nm	NIR.902 nm	NIR.904 nm	NIR.906 nm	NIR.908 nm	NIR.910 nm
1	85.3	-0.050	-0.046	-0.042	-0.037	-0.033	-0.031
2	85.25	-0.044	-0.040	-0.036	-0.031	-0.027	-0.024
3	88.45	-0.047	-0.041	-0.037	-0.031	-0.027	-0.023
4	83.4	-0.047	-0.042	-0.039	-0.035	-0.030	-0.028
5	87.9	-0.051	-0.045	-0.041	-0.036	-0.033	-0.031
6	85.5	-0.048	-0.043	-0.039	-0.034	-0.030	-0.028
7	88.9	-0.050	-0.045	-0.041	-0.036	-0.032	-0.030
8	88.3	-0.049	-0.044	-0.039	-0.034	-0.030	-0.026
9	88.7	-0.050	-0.044	-0.040	-0.035	-0.031	-0.028
10	88.45	-0.051	-0.046	-0.042	-0.037	-0.033	-0.030
11	88.75	-0.053	-0.048	-0.044	-0.039	-0.036	-0.034
12	88.25	-0.050	-0.045	-0.041	-0.036	-0.032	-0.030
13	87.3	-0.048	-0.044	-0.039	-0.034	-0.030	-0.027
14	88	-0.047	-0.041	-0.037	-0.031	-0.027	-0.024
15	88.7	-0.042	-0.037	-0.032	-0.027	-0.021	-0.018
16	85.5	-0.049	-0.044	-0.040	-0.036	-0.032	-0.029
17	88.65	-0.052	-0.046	-0.043	-0.038	-0.034	-0.031
18	88.75	-0.055	-0.050	-0.046	-0.041	-0.036	-0.034
19	85.4	-0.055	-0.049	-0.046	-0.041	-0.037	-0.034
20	88.6	-0.054	-0.048	-0.045	-0.040	-0.035	-0.032
21	87	-0.056	-0.052	-0.048	-0.043	-0.039	-0.037
22	87.15	-0.042	-0.037	-0.033	-0.028	-0.024	-0.022
23	87.05	-0.056	-0.051	-0.047	-0.042	-0.037	-0.035
24	87.25	-0.057	-0.051	-0.047	-0.042	-0.038	-0.035
25	86.85	-0.057	-0.051	-0.047	-0.043	-0.038	-0.036

The rows in the data table have been colored by the values of Octane. This was done by selecting **Rows > Color or Mark by Column**. We used the default continuous intensity scale (**Colors > Blue to Gray to Red**), where high values of octane rating are shown in red and low values in blue.

Creating a Test Set Indicator Column

A column called Test Set has been added to the data table. This column shows which rows will be used in building a model (indicated by values of "0") and which will later be used to test the model (indicated by values of "1").

Although we use the Test Set column that is already in the data table, follow these instructions to see how you would create such a column.

1. Create a new column. Select **Cols > New Column**. (You can also do this by double-clicking in the column heading area to the right of the last column in the data table, and then right-clicking in that area and selecting **Column Info**.)

2. In the New Column window, enter a **Column Name**.

3. Set the **Modeling Type** to **Nominal**.

108 *Discovering Partial Least Squares with JMP*

4. From the **Initialize Data** list, select **Random**. This provides the options shown in Figure 6.2.

5. Select **Random Indicator**. We accepted the default 80% and 20% split into 0s and 1s.

6. Click **OK**.

Figure 6.2: New Column Window for Defining Training and Test Sets

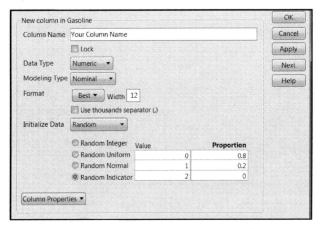

This creates a new column that you can use to select your training and test sets. However, for the remainder of this chapter, it would be best if you used the column Test Set that is already saved in the data table. So you should now delete the column that you have just created. To do this, right-click on the column heading and select **Delete Columns** from the menu.

Viewing the Data

Octane and the Test Set

Run the first script in the table, Distribution of Octane and Test Set, to see the distributions for Octane and the Test Set indicator column. The plot for Octane shows that the octane ratings range between about 83 and 90. The Test Set distribution shows two bars, because we have assigned it the **Nominal** modeling type. As expected, we have an 80/20 percent split into 0s and 1s. Click on the bar labeled "1" to see the distribution of Octane values in the test set (Figure 6.3). This suggests that we have a fairly representative selection of Octane values in the test data.

Figure 6.3: Distribution of Octane Values in the Test Set

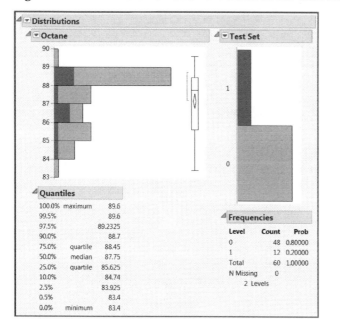

Creating a Stacked Data Table

JMP provides many ways to view spectral data. We will stack the data into a new data table and then create overlay plots to view the individual spectra as well as all of the spectra together. The script **Stack NIR Columns** creates a new data table where all spectral readings are stacked in a single column. This script also places a script in the new data table, called **Plots of Individual Spectra**. This script creates the overlay plots whose construction is described in the next section, "Constructing Plots of the Individual Spectra".

You can run the script **Stack NIR Columns** or you can follow these instructions to first create the data table and then plot the individual spectra yourself using menu commands.

With the data table Gasoline.jmp as the active data table, complete the following steps:

1. Select **Tables > Stack**.

2. In the text box to the right of **Output table name**, enter a name for the data table. We have called ours **Stacked Data**.

3. Select the NIR Wavelengths column group and click **Stack Columns**.

4. In the text box to the right of **Stacked Data Column**, give the stacked reflectance values the column name Reflectance.

5. In the text box to the right of **Source Label Column**, give the source columns the name Wavelength.

6. Figure 6.4 shows the populated Stack window. Click **OK**.

Figure 6.4: Populated Stack Window

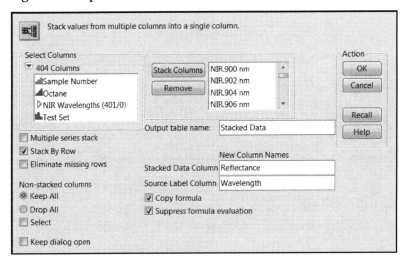

The stacked data table is partially shown in Figure 6.5. Note that the row state colors are inherited by the stacked data.

Figure 6.5: Partial View of Stacked Data Table

Sample Number	Octane	Test Set	Wavelength	Reflectance
1	85.3	0	NIR.900 nm	-0.050
2	85.3	0	NIR.902 nm	-0.046
3	85.3	0	NIR.904 nm	-0.042
4	85.3	0	NIR.906 nm	-0.037
5	85.3	0	NIR.908 nm	-0.033
6	85.3	0	NIR.910 nm	-0.031
7	85.3	0	NIR.912 nm	-0.030
8	85.3	0	NIR.914 nm	-0.031
9	85.3	0	NIR.916 nm	-0.034
10	85.3	0	NIR.918 nm	-0.036
11	85.3	0	NIR.920 nm	-0.040
12	85.3	0	NIR.922 nm	-0.043
13	85.3	0	NIR.924 nm	-0.047
14	85.3	0	NIR.926 nm	-0.048
15	85.3	0	NIR.928 nm	-0.051
16	85.3	0	NIR.930 nm	-0.054
17	85.3	0	NIR.932 nm	-0.055
18	85.3	0	NIR.934 nm	-0.057
19	85.3	0	NIR.936 nm	-0.058
20	85.3	0	NIR.938 nm	-0.061

All rows: 24,060

Constructing Plots of the Individual Spectra

With the stacked data table that you have just created as the active data table, complete the following steps. (If you don't want to construct these plots using the menu options, you can simply run the script Stack NIR Columns in Gasoline.jmp and then run the script that appears in the stacked data table.)

1. Select **Graph > Overlay Plot**.

2. Select Reflectance and click **Y**.

3. Select Wavelength and click **X**.

4. Select Sample Number and click **By**.

5. Deselect the **Sort X** box under **Options**.

   Deselecting the **Sort X** option allows the Wavelength values to appear on the horizontal axis in the order in which they appear in the data table. Otherwise, these would be sorted alphanumerically.

   Figure 6.6 shows the populated Overlay Plot window.

6. Click **OK**.

Figure 6.6: Populated Overlay Plot Window

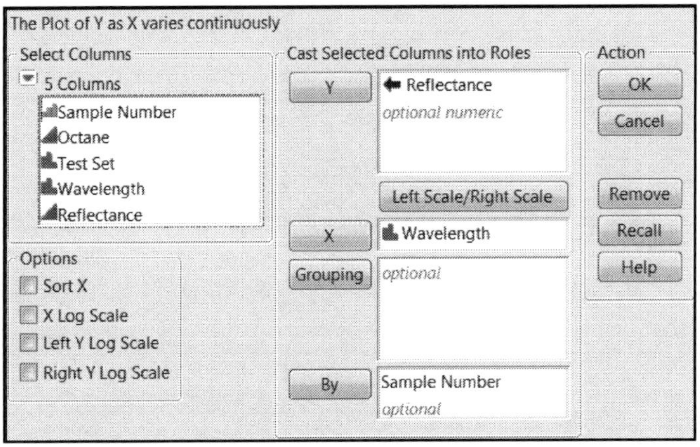

Individual Spectra

A partial view of the individual spectra is given in Figure 6.7. Recall that the color of the points describes the Octane value, with red indicating higher values and blue indicating lower values. As you scroll through these individual spectra, you notice a common "fingerprint" consisting of two peaks followed by an increase in reflectance for the longer wavelengths.

Had you wanted to see the Test Set designation for each sample, you could have added Test Set as a **By** variable in the Overlay Plot window.

Figure 6.7: Partial View of Overlay Plots for Individual Spectra

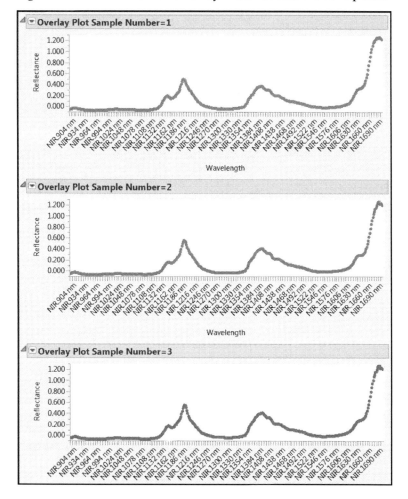

Combined Spectra

Once you have reviewed the spectra, close the stacked data table. In the data table Gasoline.jmp, run the script Review All Spectra. This script produces a plot containing the spectra for all samples, along with a **Local Data Filter** (Figure 6.8).

Because this script displays the data using the **Parallel Plot** platform (with uniform scaling), it does not require the data to be stacked. The **Local Data Filter** enables you to see one or more selected spectra in the context of the rest. The advantage of using a **Local Data Filter** rather than the (global) **Data Filter** obtained under the **Rows** menu is that it does not affect the row states in the data table. Our data table contains color row states; these are unaffected by the **Local Data Filter**.

Figure 6.8: Local Data Filter and Parallel Plot

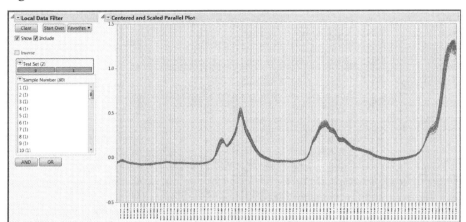

To construct this plot on your own using menu options, complete these steps. Make sure that Gasoline.jmp is your active data table.

1. Select **Graph > Parallel Plot**.
2. In the **Select Columns** list, select the NIR Wavelengths column group and click **Y, Response**.
3. Select the **Scale Uniformly** check box.

   The **Scale Uniformly** option ensures that the vertical axis represents the actual measurement units. Because parallel plots are often used to compare variables that are measured on different scales, the JMP default is to leave this check box deselected.

   The Parallel Plot window should appear as shown in Figure 6.9.
4. Click **OK**.
5. In the **Parallel Plot** report, click the red triangle and select **Script > Local Data Filter**, as shown in Figure 6.10.

   This adds a **Local Data Filter** to the **Parallel Plot** report.

6. In the **Local Data Filter** report, from the **Add Filter Columns** list, select Test Set and then click **Add**.

7. Click the **AND** button.

8. From the **Add Filter Columns** list, select Sample Number and click **Add**.

Your report now appears as shown in Figure 6.8.

Figure 6.9: Populated Parallel Plot Window

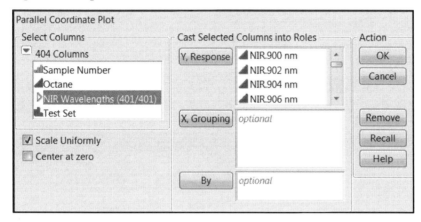

Figure 6.10: Local Data Filter Selection from Parallel Plot Report Menu

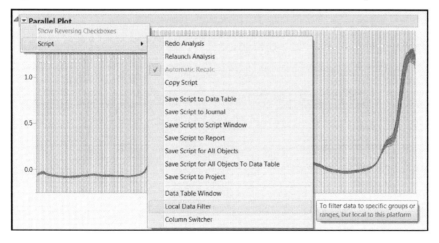

At this point, you can click on a value of **Test Set** to compare the spectra in the two sets. Or you can click on one or more values of **Sample Number** to see that row or collection of rows (hold down the Ctrl key to make multiple selections). Once you have finished exploring the data, you can close all open reports.

In this example, we have used the built-in capabilities of JMP to explore the data visually. If you do a lot of work with spectral data, you should look at the File Exchange on the JMP User Community page for add-ins that might be useful to you.

A First PLS Model

Excluding the Test Set

Now we turn our attention to developing a PLS model for **Octane**. We'll develop this model using our training set, namely, those rows for which **Test Set** has the value 0. So, at this point, **Hide** and **Exclude** all rows for which Test Set = 1. An easy way to do this is as follows (or you can run the script Exclude Test Set):

1. Select **Analyze > Distribution**.
2. Enter Test Set in the **Y, Columns** list box.
3. Click **OK**.
4. In the **Distribution** report, click on the bar labeled "1".
5. Then right-click on this bar and, from the menu, select **Row Hide and Exclude** (Figure 6.11).

This excludes and hides 12 rows in the data table. Check this in the data table **Rows** panel.

Figure 6.11: Excluding and Hiding Rows Using Distribution

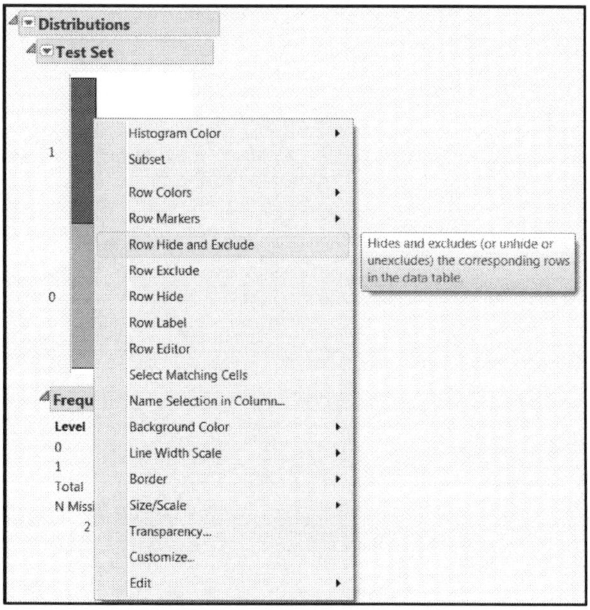

Fitting the Model

Next, we fit a PLS model using the 48 rows in the training set. We use Fit Model to obtain the PLS Model Launch control panel, giving the instructions for JMP Pro. (Alternatively, if you have JMP Pro, you can run the PLS Model Launch script.) If you have JMP, you can use the PLS launch by selecting **Analyze > Multivariate Methods > Partial Least Squares**.

1. Select **Analyze > Fit Model**.
2. From the **Select Columns** list, select Octane and click **Y**.
3. Select the NIR Wavelengths column group and click **Add**.
4. From the **Personality** menu, select **Partial Least Squares**.
5. Click **Run**.

The Partial Least Squares Model Launch control panel (Figure 6.12) opens.

Figure 6.12: PLS Model Launch Control Panel

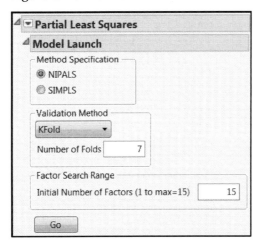

We accept the default settings. Note that we are using the default **Validation Method** (**KFold** with **Number of Folds** equal to **7**). This involves a random assignment of rows to folds. In order for you to reproduce the results shown in our analysis, we suggest that you do the following. From the red triangle menu, select **Set Random Seed**. In the window that opens, enter the value **666** as the random seed and click **OK**.

If you are using JMP, use **Leave-One-Out** as the **Validation Method**. Be aware that your results will differ from those obtained below.

The Initial Report

Click **Go** in the Model Launch control panel. The initial reports include a **Model Comparison Summary** table, a **KFold Cross Validation** report, and a **NIPALS Fit with 5 Factors** report (Figure 6.13).

Figure 6.13: Model Fit Results

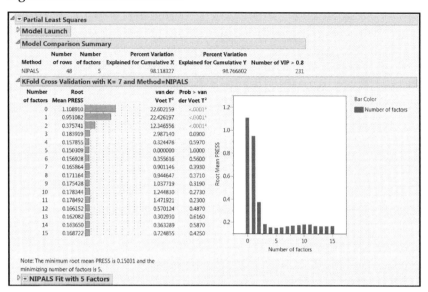

The **KFold Cross Validation** report computes the **Root Mean PRESS** statistic, using the seven folds, for models based on up to 15 factors. Note that this requires fitting 7 x 15 = 105 PLS models. (Details about the calculation are given in "Determining the Number of Factors" in Appendix 1.) The report shows that the **Root Mean PRESS** statistic is minimized using a five-factor model. This is the initial model that JMP fits. Details for this model are given in the **NIPALS Fit with 5 Factors** report.

The **Model Comparison Summary** gives high-level information about the five-factor model that has been fit. It notes that 48 rows were used to construct this model and that the model explains 98.12% of the variation in the **X**s and 98.77% of the variation in **Y**. The **Model Comparison Summary** table is updated when new models are fit.

The **KFold Cross Validation** report also provides test statistic values (**van der Voet T^2**) and p-values (**Prob > van der Voet T^2**) for a test developed by van der Voet (1994). For a model with a given number of factors, the van der Voet test compares its PRESS statistic with the PRESS statistic for a model based on the number of factors corresponding to the minimum PRESS value achieved. This enables one to determine whether the variation explained by the additional factors is significant. A cut-off of 0.10 for **Prob > van der Voet T^2** is often used.

The first three **Prob > van der Voet T²** values in the **KFold Cross Validation** report shown in Figure 6.13 are displayed in red and have asterisks to their right. They are all less than 0.0001. This indicates that the residuals from the null, one-, and two-factor models differ significantly from those for the five-factor model. However, residuals from the four-factor model do not differ significantly from those of the five-factor model. Although the *p*-value for the three-factor model is slightly below 0.10, the **Root Mean PRESS** plot to the right of the *p*-values suggests that it is similar to the five-factor model. This leads us to select the three-factor model as our model of choice.

A Second PLS Model

Fitting the Model

To fit the three-factor model, in the Model Launch control panel, enter **3** as the **Number of Factors** in the **Factor Specification** text box, and click **Go**. The **Model Comparison Summary** report is updated accordingly, and a report for the new fit, **NIPALS Fit with 3 Factors**, is added to the report window (Figure 6.14).

Figure 6.14: Three-Factor Model

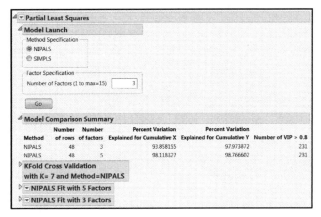

High-Level Overview

For our candidate three-factor model, the percent variation in **Y** that is explained is about 98%, while the percent variation in **X** that is explained is about 94%.

Let's examine the reports provided as part of the **NIPALS Fit with 3 Factors** report (Figure 6.15). We'll start by reviewing the **X-Y Scores Plots** and the **Percent Variation Explained** report.

Figure 6.15: X-Y Scores Plots and Percent Variation Explained Report

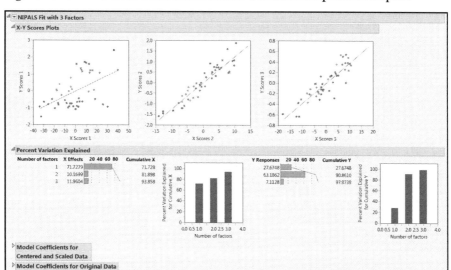

In the **X-Y Scores Plots**, we hope to see a lot of the variation in the Y scores explained by the X scores. For each factor, the Y scores are plotted against the X scores and a least squares line is added to describe the relationship. In this case, the first factor explains some variation in **Y**, but there is scatter about the line. The second and third factors explain more variation, as seen by the fact that the points are more tightly grouped around the line. Note also that, compared to the first factor, the second factor provides much better discrimination between the measured Octane rating, with red points to the upper right and blue points to the lower left.

The behavior exhibited by the scores plots is related to the values presented in the **Percent Variation Explained** report. The first factor explains about 72% of the **X** variation, while it explains only about 28% of the **Y** variation. These roles are switched for the second factor, which explains only 10% of the variation in **X** but 63% of the variation in **Y**. Because it explains 72% of the variation in **X**, the first factor represents the observed values of the predictors well, while the second factor represents the values of **Y** fairly well.

Diagnostics

Although our model appears to explain a lot of variation, before we accept it as a good model, we need to scrutinize it for shortcomings. So, before proceeding to a detailed analysis of the scores, loadings, VIP values, and model coefficients, let's look at a few diagnostic tools.

Select **Diagnostic Plots** from the red triangle menu for the **NIPALS Fit with 3 Factors** report. This gives the four plots shown in Figure 6.16. The **Actual by Predicted Plot** and the **Residual by Predicted Plot** show evidence that residuals increase with larger predicted values of Octane. This is not unexpected, and it is not serious enough to affect our ability to fit a predictive model. Had this been more pronounced, we might have considered using a normalizing transformation for Octane, rather than the raw Octane value itself. The **Residual by Row Plot** and the **Residual Normal Quantile Plot** give no cause for concern.

Figure 6.16: Diagnostics Plots

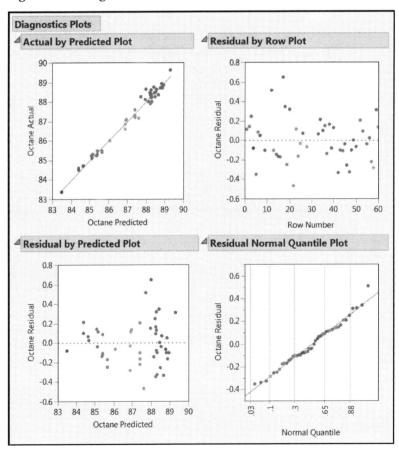

Next, select **Distance Plots** and **T Square Plot** from the red triangle menu for the PLS fit (Figure 6.17). The first two **Distance Plots** show distances from each of the observations to the X model and the Y model. The third plot shows an observation's distance to the Y model plotted against its distance to the X model. None of the points stand out as unusually distant from the X or Y models.

The **T² Plot with Control Limit** shows a multivariate measure of the distance from each point's X scores on the three factors to their mean of zero. A control limit, whose value is based on the assumption of multivariate normality of the scores, is plotted. By positioning your mouse pointer over the point that falls above the limit, you obtain the tooltip indicating that this is observation 15.

Figure 6.17: Distance and T Square Plots

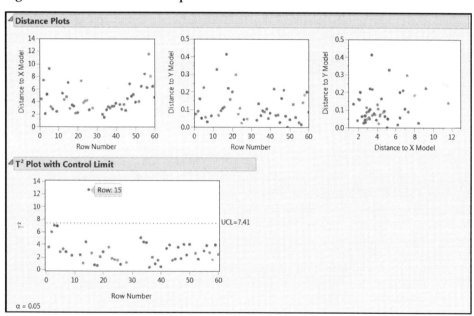

Click to select this point. Then right-click in the plot and select **Row Label**. This labels the observation by its row number, namely, 15. Click on some blank space in the plot to deselect observation 15. Now look at the **X-Y Scores Plots** (Figure 6.18). Observation 15's X scores are among the more extreme X scores. The X score for observation 15 is high for factors 1 and 2 and low for factor 3.

Figure 6.18: X-Y Scores Plots with Observation 15 Labeled

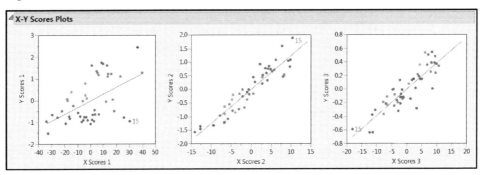

Looking in the data table, you can see that the Octane rating for observation 15 is 88.7, among the highest Octane ratings in our training set. You can select **Analyze > Distribution** to verify this.

To get some insight on why observation 15 is extreme relative to the X scores, we'll do the following: Run the script Review All Spectra. Take note of where the low (bluish) Octane colors appear. Now select **Sample Number 15** in the **Local Data Filter**. Click on the trace for observation 15. Deselect **Show**. The spectral trace for observation 15 appears highlighted on the parallel plot, but the other traces appear as well (Figure 6.19). Notice that observation 15 tends to be extreme in terms of its reflectance values—its trace is usually at one or the other extreme of the other traces. Also, its trace is unusual given its Octane value. Over the spectrum, its trace often falls in largely "blue," rather than "red," regions.

Figure 6.19: Spectral Trace for Observation 15

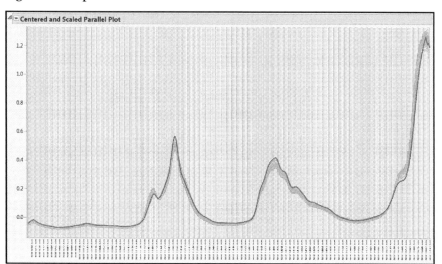

It is likely that this behavior is causing the X scores for observation 15 to be extreme, and that this is what is being described by the **T² Plot with Control Limit**. In terms of distance to the X and Y models, observation 15 is not aberrant. But the **T² Plot with Control Limit** indicates that observation 15 is somewhat distant in the factor-defined predictor space.

If this is a legitimate and correct observation, then one should retain it in the model. As is often the case in the MLR setting, an observation might have high leverage, but if it is a legitimate observation, it might still be reasonable to retain it in the model. If this were your data, you might want to verify that observation 15 was not affected by unusual circumstances.

We proceed with observation 15 as part of our training set. However, if you desired, it would be easy for you to exclude this observation and refit the model. At this point, remove the label from observation 15. You can either do this by right-clicking on the point in a plot and selecting **Row Label** or by going back to the data table, right-clicking on the label icon, and selecting **Label/Unlabel**.

Score Scatterplot Matrices

You can also plot the X scores or Y scores against each other to look for irregularities in the data. To do this, select **Score Scatterplot Matrices** from the red triangle menu next to the **NIPALS Fit with 3 Factors** report title. In the plots, you should look for patterns or clusters of observations. For example, if you see a curved pattern, you might want to add a quadratic term to the model. Two or more clusters of observations indicate that it might be better to analyze the groups separately.

The **X Score Scatterplot Matrix** is shown in Figure 6.20. Each scatterplot displays a 95% confidence ellipse. For insight on how the ellipses are constructed, think of X Score i as plotted on the horizontal axis and X Score j as plotted on the vertical axis. The pairs of scores, (X Score i, X Score j), are assumed to come from a bivariate normal distribution with zero covariance. The assumption of zero covariance follows from the orthogonality of the X scores, which results from the process of deflation. (See Chapter 4, "The NIPALS and SIMPLS Algorithms" and Appendix 1, "Properties of the NIPALS Algorithm.") A T^2 statistic is computed for the X scores (Nomikos and MacGregor 1995, and Kourti and MacGregor 1996). The exact distribution of this statistic, times a constant, is a beta distribution (Tracy et al. 1992).

Figure 6.20: X Score Scatterplot Matrix

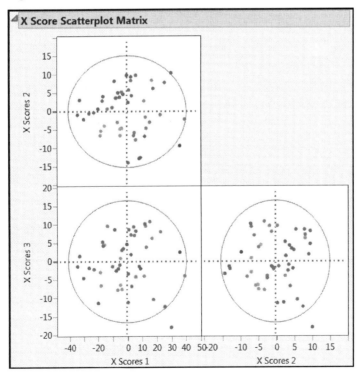

Looking at the **X Score Scatterplot Matrix**, we see no problematic patterns or groupings. You can think of the X-score plots as describing the coverage or support of the study, in terms of the factors. In that vein, the data seem fairly well dispersed through the factor space. Because each X score is a linear combination of the 401 predictors, the bivariate normality assumption required for the interpretability of the ellipses is viable. By positioning your mouse pointer over the points that fall just outside the ellipses, you see that these points correspond to observations 4 and 15. Observation 4 is hardly a cause for concern. We have already seen why observation 15 has extreme values for X scores. Although it would be prudent to verify that observation 15 is not suspect or different in some sense, there is nothing in the plots to suggest that it poses a serious issue.

The color coding applied to Octane helps us see that X Scores 1 does not explain Octane values well. However, X Scores 2 seems to distinguish high from low Octane values. This is consistent with the **Percent Variation Explained** report, which indicates that the first factor tends to explain variation in **X**, rather than **Y**, while the second factor explains a large proportion of the variation in **Y**.

Because, in this example, **Y** consists of a single vector, the **Y Score Scatterplot Matrix** gives information about deflation residuals. More specifically, because there is a single **Y**, it follows that Y Scores 1 is collinear with Octane. Because we are using the NIPALS method to fit our model, Y Scores 2 consists of the residuals once Y Scores 1 (which is collinear with Octane) has been regressed on X Scores 1, and Y Scores 3 consists of the residuals once Y Scores 2 has been regressed on X Scores 2. (To be precise, these relationships hold up to ±1.) So Y Scores 2 consists of the variation in Octane that the first factor fails to predict, and Y Scores 3 consists of the variation that the first two factors fail to predict.

Loading Plots

Next, let's look at the loadings. The loadings for a factor indicate the degree to which each wavelength is correlated with, or contributes to, that factor. There are three options for viewing loadings: **Loading Plots**, **Loading Scatterplot Matrices**, and **Correlation Loading Plot**. For spectral data, the best option is **Loading Plots**, because this presentation uses overlay plots and preserves the ordering of the wavelengths.

Select **Loading Plots** from the red triangle menu for the three-factor fit. Note that the check boxes to the left of the plot allow you to select which factor loadings to display. (See Figure 6.21, where loadings for all three factors are displayed.)

Figure 6.21: Loading Plots

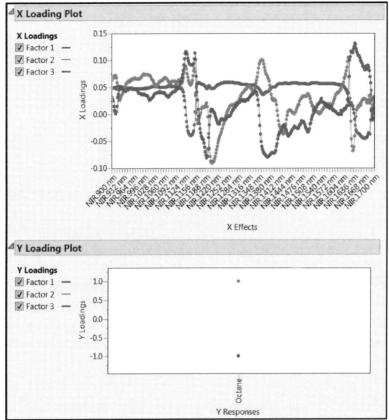

The **Y Loading Plot** is not of interest because there is only one **Y** and each loading value must be ±1. The **X Loading Plot** shows that features in different spectral regions seem to characterize the three factors. These factors define the "fingerprints" that we mentioned in "Why Use PLS?" in Chapter 4. Factor 1 seems to be weakly, but positively, correlated with most wavelengths, except in two notable areas of the spectrum, where ranges of wavelengths are negatively correlated with factor 1. Factors 2 and 3 have interesting loading patterns: Factor 2 seems to have three dips and two peaks; Factor 3 seems to have two dips and three peaks.

Also, the **X Loading Plot** shows that there are relatively few variables with loadings near zero, suggesting that most NIR measurements contribute to defining the factors.

VIPs

Recall that PLS factors are defined so as to both explain variation in **X** and **Y** and to build a useful regression model. There are various reasons why we might be interested in identifying the variables that contribute substantively to the model. It might be useful to know that certain variables don't contribute to help better understand the mechanism at work, or simply to avoid wasting time in the future in measuring unhelpful variables. Also, if certain variables only contribute noise to the model, a better and more parsimonious model can be obtained by eliminating these variables.

A variable's VIP (variable importance for the projection) value measures its influence on the factors that define the model. To see the VIPs for our 401 wavelengths, select **Variable Importance Plot** from the red triangle menu at the **NIPALS Fit with 3 Factors** report title (Figure 6.22). The dashed horizontal line in the plot is set at the threshold value of 0.8. This is the cut-off value advocated by Wold (1995, p. 213) to separate terms that do not make an important contribution to the dimensionality reduction involved in PLS (VIP < 0.8) from those that might (VIP ≥ 0.8). We already know from the **Model Comparison Summary** that 231 of the 401 wavelengths have VIP values exceeding 0.8.

Figure 6.22: Variable Importance Plot

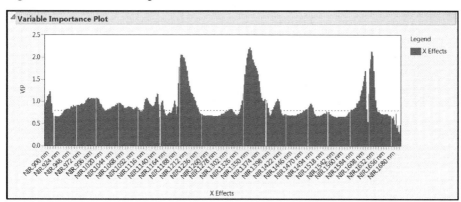

Compare the **X Loading Plot** in Figure 6.21 to the **Variable Importance Plot** in Figure 6.22. The spectral regions with high VIP values seem to mirror the regions where the factor loadings distinguish the factors.

The report below the plot, **Variable Importance Table**, gives the VIP values for the variables in a form that can be made into a data table. To construct such a data table, click on the disclosure icon to reveal the table and its associated bar graph, and then right-click in the report and select **Make into Data Table**.

The VIP value does not directly reflect a variable's role in the overall predictive model. That variable's regression coefficient, based on the centered and scaled data, gives a way to determine its influence on the predictive model for **Y**. As mentioned in Chapter 5, common practice is to consider variables with VIPs below 0.8 and with standardized regression coefficients near 0 to be unimportant relative to the overall PLS model.

From the red triangle menu at **NIPALS Fit with 3 Factors**, select **VIP vs Coefficients Plots**. If the **Centering** and **Scaling** options have been selected in the PLS launch window, the **VIP vs Coefficients Plots** option gives two plots: one for the centered and scaled data; one for the original data (with Xs standardized if the **Standardize X** option was selected).

Regression coefficients are sensitive to centering and scaling. For this reason, we consider the plot for the centered and scaled data. (See Figure 6.23, where we have deselected the **Show Labels** options.) The plot gives VIP values on the vertical axis and regression coefficient values on the horizontal axis.

Figure 6.23: VIP versus Coefficients for Centered and Scaled Data

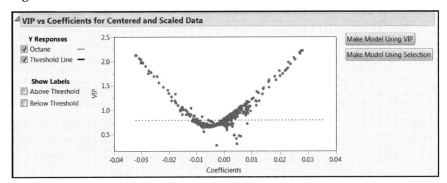

The dashed horizontal line is set at the Wold threshold value of 0.8. With both VIPs and coefficients displayed in a single plot, it is easier to make decisions about which variables to consider unimportant relative to the PLS model. The **Make Model Using VIP** button creates a model containing only those variables with VIPs exceeding 0.8. The **Make Model Using Selection** button constructs a model based on points that you choose interactively in the plot.

The **VIP vs Coefficients for Centered and Scaled Data** plot often exhibits a characteristic "V" shape. In particular, this suggests that variables with low VIP values also tend to have small values for their standardized regression coefficients. If one were to define a new model using only those variables with VIPs exceeding 0.8, the excluded variables would generally have small magnitude regression coefficients as well.

But note that there is one variable with VIP below the 0.8 threshold that has a regression coefficient value near 0.01. To identify that variable, position your mouse pointer over it in the plot. We have selected it in Figure 6.24. It is NIR.1690 nm, when the wavelength is 1690 nanometers. This wavelength is near the upper end of the measured spectrum. If you look closely at Figure 6.22, you can spot it by its relatively large VIP value. We revisit this wavelength later on.

Figure 6.24: Identifying Odd Wavelength (Coefficient about 0.009 and VIP about 0.73)

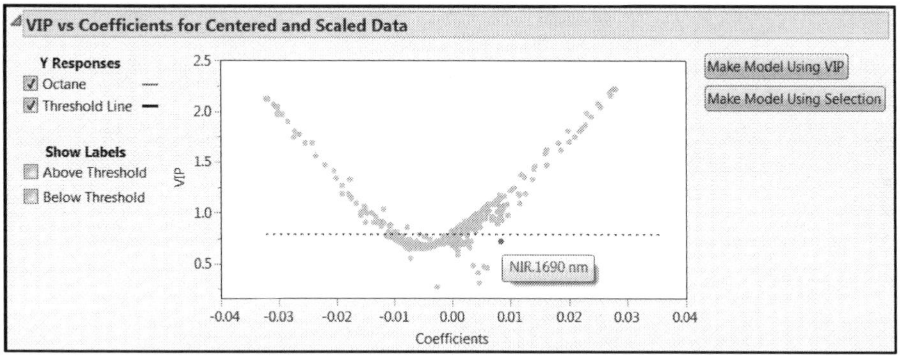

Viewing VIPs and Regression Coefficients for Spectral Data

As is the case for the loadings, for spectral data it is useful to see VIP values and coefficients in a plot that preserves the ordering of wavelengths. We illustrate a method for doing this.

If you have closed your **Variable Importance Plot**, please reselect it. Also, select **Coefficient Plots** from the red triangle menu of the PLS fit. We show how to use these plots and JMP journal abilities to create the plot shown in Figure 6.25. (The completed journal file is called VIPandCoefficients.jrn. You can open this file by clicking on the VIPandCoefficients link in the master journal.)

Figure 6.25: Variable Importance and Coefficient Plots with Graphics Scripts

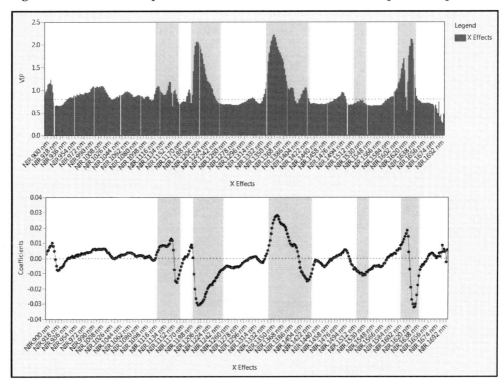

Select **File > New > Journal** to make a new journal file and to avoid appending content to the open master journal. In the report, use the "fat plus" tool and hold the Shift key to select the **Variable Importance Plot** and the **Coefficient Plot for Centered and Scaled Data**. With the right selection you should be able to include the graphics region and the axes and labels, but exclude the legends. Now select **Edit > Journal** to copy the selected regions to the journal file. Exchange the "fat plus" tool for the pointer, and resize the two graphics boxes to make the wavelength axes align properly. (For information about journal files, select **Help > Books > Using JMP** and search for "JMP Journals".)

In Figure 6.25, we have added graphics scripts to each of the plots to identify wavelength ranges (using green) that seem to be of special interest relative to modeling **Octane**. These ranges could also form the basis for a more parsimonious model. We have generally followed the idea that the interesting wavelengths have VIP values exceeding

0.8 or standardized regression coefficients that are not near zero. We later reduce our model using the 0.8 VIP cut-off, but for now we simply want to illustrate how spectral ranges of interest might be identified.

To add a graphics script, right-click in either plot and select **Customize**. In the Customize Graph window, click the **+** sign button. You can add a graphics script in the text window. To view the graphics scripts we created, select the journal file, VIPandCoefficients.jrn, right-click in a plot, select **Customize**, and click **Script** in the list. Close the journal file you have constructed.

Model Assessment Using Test Set

We have analyzed our three-factor model and have found no reason not to consider it. Given that the goal of our modeling effort is prediction, it is time to see how well our candidate three-factor PLS model performs, particularly on the test set. We will display predicted versus actual Octane values on a scatterplot for both our training and test sets.

First we'll save the prediction formula for Octane. To accomplish this, complete the following steps, or run the script called Save Prediction Formula.

1. From the **NIPALS Fit with 3 Factors** red triangle menu, select **Save Columns > Save Prediction Formula**. (Or, run the script Save Prediction Formula.)

    This adds a new column called Pred Formula Octane to the data table.

    Next, we'll apply a marker to the test set and create a plot of predicted versus actual values (Figure 6.26). Alternatively, you can run the script Predicted versus Actual Octane.

2. Select **Rows > Clear Row States**.

3. Reapply the Octane colors by selecting **Rows > Color or Mark by Column**, and selecting Octane in the Mark by Column window. (Or, run the script Color by Octane.)

4. Select the rows for which Test Set equals 1. An easy way to do this is to select **Analyze > Distribution**, enter Test Set as **Y, Columns**. In the report's bar graph, click in the bar corresponding to Test Set = 1. This selects the 12 rows in the test set.

5. Select **Rows > Markers** and click on the open circle, **o**, to use that symbol to identify test set observations.

6. In the upper left corner of the data table, above the row numbers, click in the lower triangular region to clear the row selections.

7. Select **Graph > Graph Builder**.

8. Drag Pred Formula Octane to the **Y** axis drop zone. Drag Octane to the **X** axis drop zone.

9. Click on the second icon from the left above the graph to deselect the **Smoother**.

If you want to increase the size of the points to match ours, right-click in the plot, select **Graph > Marker Size**, and then select **4, XL**.

Your plot should look like the one in Figure 6.26 (where we have deselected the **Show Control Panel** option in the **Graph Builder** red triangle menu). The actual values and predicted values align well. Also, we don't see any systematic difference in the fit between the training and test set observations.

Figure 6.26: Plot of Predicted versus Actual Octane Values

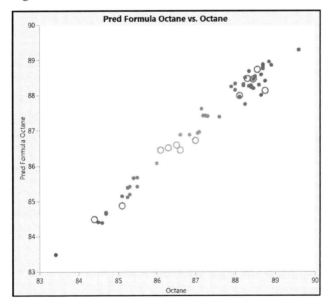

To get a better look at our prediction error, we examine the residuals. Note that residuals can be saved to a column by selecting **Save Columns > Save Y Residuals**, but this option does not save a formula. So, we create a column containing the formula for residuals. (The script is Create Residuals Column.)

1. Select **Cols > New Column** to create a new column.
2. Enter Residuals as the **Column Name**.
3. From the **Column Properties** menu, select **Formula**.
4. Enter the formula **Octane – Pred Formula Octane**.
5. Click **OK**.

    Next, we create a plot using **Graph Builder**. Alternatively, you can run the script Residuals versus Octane.

6. Select **Graph > Graph Builder**. Drag Residuals to the **Y** axis drop zone. Drag Octane to the **X** axis drop zone.
7. Click on the second icon from the left above the graph to deselect the **Smoother**.
8. Double-click the Residuals axis to open the Y Axis Specification window.
9. Click **Add** to insert a reference line at **0**.
10. Click **OK**.

If you want to increase the size of the points to match ours, right-click in the plot, select **Graph > Marker Size**, and then select **4, XL**.

The plot is shown in Figure 6.27 (where, again, we have deselected the **Show Control Panel** option in the **Graph Builder** red triangle menu). The plot does not show any systematic difference in the fit between the training and test set observations.

From the plot, we see that the largest value of prediction error for our test set is a little over 0.6 units. Given that the measured sample octane ratings range from 83 to 90, this amount of prediction error seems small, and indicates that we have a model that should be of practical use.

Figure 6.27: Residual Plot

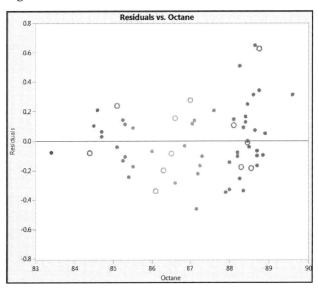

A Pruned Model

At this point, we wonder if a model that eliminates variables that don't contribute substantially might prove as effective as the model we have just developed. We fit a new model that includes only those variables with VIPs of 0.8 and above, and NIR.1690 nm.

To fit this model, complete the following steps. (The script is PLS Pruned Model Launch.)

1. In the data table, run the script Exclude Test Set to re-exclude the test observations.

2. In the **NIPALS Fit with 3 Factors** red triangle menu, select **VIP versus Coefficients Plots**.

3. In the **VIP vs Coefficients Plot for Centered and Scaled Data** report, to the right of the plot, click **Make Model Using VIP**.

    This enters all variables with VIPs exceeding 0.8 into a Fit Model or PLS launch window.

4. In the launch window, in the **Select Columns** list, open the NIR Wavelengths column group and select NIR.1690 nm.

Recall that this was the wavelength that had a VIP close to 0.8 and a comparatively large regression coefficient.

5. Add NIR.1690 nm to the list of model effects. (In the Fit Model launch window, click **Add**; in the PLS launch window, click **X, Factor**.)

6. Click **Run** (Fit Model launch window) or **OK** (PLS launch window).

7. To obtain the same JMP Pro results as we discuss in the report below, click the red triangle, select **Set Random Seed**, set the value of the seed to **666**, and click **OK**.

8. Accept the default settings and click **Go**.

The van der Voet test in the report **KFold Cross Validation with K = 7 and Method = NIPALS** suggests that a four-factor model does not differ statistically from the minimizing five-factor model. Return to the Model Launch control panel and enter **4** as the **Number of Factors** under **Factor Specification**. Then click **Go** to fit this model. (The script is Fit Second Pruned Model.)

The **Percent Variation Explained** report (Figure 6.28) shows that the four factors explain about 98% of the variation both in **X** and in **Y**. This is better than our previous three-factor model. Even the first three factors in this pruned model explain slightly more variation than did our previous model.

Figure 6.28: Percent Variation Explained Report for Pruned Model

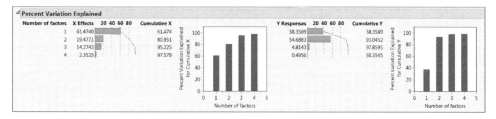

We leave it as an exercise for you to save the prediction formula and to construct a column containing residuals. You can easily adapt the instructions in the section "Model Assessment Using Test Set". Also, following those instructions, you can construct a plot of residual versus actual Octane values, as shown in Figure 6.29. (The script is Residuals Versus Actual for Pruned Model.)

Figure 6.29: Residual Plot for Pruned Model

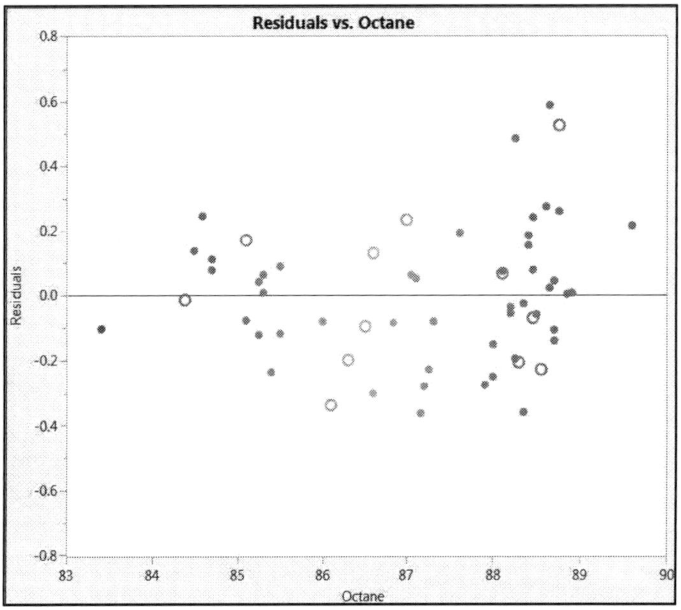

We conclude that this model fits well and might even be superior to the original model. In addition, the model might have identified spectral ranges that are of scientific interest.

Water Quality in the Savannah River Basin

Background 140
The Data 141
    Data Table Description 141
    Initial Data Visualization 144
    Missing Response Values 145
    Impute Missing Data 146
    Distributions 147
    Transforming AGPT 148
    Differences by Ecoregion 150
    Conclusions from Visual Analysis and Implications 155
A First PLS Model for the Savannah River Basin 155
    Our Plan 155
    Performing the Analysis 156
    The Partial Least Squares Report 158
    The NIPALS Fit Report 159
    Defining a Pruned Model 163
A Pruned PLS Model for the Savannah River Basin 166
    Model Fit 166
    Diagnostics 168
    Saving the Prediction Formulas 169
    Comparing Actual Values to Predicted Values for the Test Set 170
A First PLS Model for the Blue Ridge Ecoregion 173
    Making the Subset 173
    Reviewing the Data 174
    Performing the Analysis 175
    The NIPALS Fit Report 176

A Pruned PLS Model for the Blue Ridge Ecoregion 178
   Model Fit 178
   Comparing Actual Values to Predicted Values for the Test Set 179
Conclusion 181

Background

The term *water quality* refers to the biological, chemical, and physical conditions of a body of water, and is a measure of its ability to support beneficial uses. Of particular interest to landscape ecologists is the relationship between landscape conditions and indicators of water quality. (See the U.S. EPA Landscape Ecology website.)

In their attempts to develop statistically valid predictive models that relate landscape conditions and water quality indicators, landscape ecologists often find themselves with a small number of observations and a large number of highly correlated predictors. An additional challenge relates to the low level of signal relative to noise inherent in the relationship between predictors and responses. In the application of standard multiple or multivariate regression analyses, these conditions usually compromise the modeling process in one way or another, often requiring the selection and use of a subset of the potential predictors.

In this chapter, you apply PLS to model the relationships among biotic indicators of surface water quality (the **Y**s) and landscape conditions (the **X**s), thus ameliorating these problems. Our example is based on the paper by Nash and Chaloud (2011). We would like to thank the authors for providing us with the data used in their paper.

The values of the **Y**s are derived from measurements made on samples of water taken at specific locations in the Savannah River basin, and the values of the **X**s are calculated from existing remote sensing data representing the associated landscape conditions in the vicinity. The remote sensing data was obtained from several sources. (For more information, see Nash and Chaloud 2011.)

Given that the investment to secure remote sensing data from wide areas of the United States has already been made, gathering new values for these **X**s is inexpensive and simple, whereas for the **Y**s such values would be expensive and time-consuming. So the promise of a viable predictive model that makes good use of available landscape data is appealing.

The Data

Data Table Description

Open the data table WaterQuality1.jmp, partially shown in Figure 7.1, by clicking on the correct link in the master journal. This table contains 86 rows, one for each sample of water taken at a specific location in the field. For each sampling location, the watershed support area was delineated and a suite of landscape variables was calculated.

Figure 7.1: Partial View of WaterQuality1.jmp

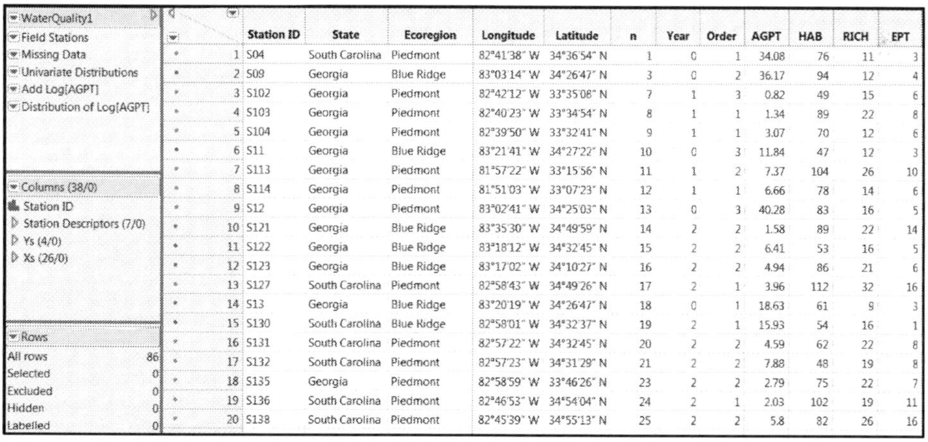

	Station ID	State	Ecoregion	Longitude	Latitude	n	Year	Order	AGPT	HAB	RICH	EPT
1	S04	South Carolina	Piedmont	82°41'38" W	34°36'54" N	1	0	1	34.08	76	11	3
2	S09	Georgia	Blue Ridge	83°03'14" W	34°26'47" N	3	0	2	36.17	94	12	4
3	S102	Georgia	Piedmont	82°42'12" W	33°35'08" N	7	1	3	0.82	49	15	6
4	S103	Georgia	Piedmont	82°40'23" W	33°34'54" N	8	1	1	1.34	89	22	8
5	S104	Georgia	Piedmont	82°39'50" W	33°32'41" N	9	1	1	3.07	70	12	6
6	S11	Georgia	Blue Ridge	83°21'41" W	34°27'22" N	10	0	3	11.84	47	12	3
7	S113	Georgia	Piedmont	81°57'22" W	33°15'56" N	11	1	2	7.37	104	26	10
8	S114	Georgia	Piedmont	81°51'03" W	33°07'23" N	12	1	1	6.66	78	14	6
9	S12	Georgia	Piedmont	83°02'41" W	34°25'03" N	13	0	3	40.28	83	16	5
10	S121	Georgia	Blue Ridge	83°35'30" W	34°49'59" N	14	2	2	1.58	89	22	14
11	S122	Georgia	Blue Ridge	83°18'12" W	34°32'45" N	15	2	2	6.41	53	16	5
12	S123	Georgia	Blue Ridge	83°17'02" W	34°10'27" N	16	2	2	4.94	86	21	6
13	S127	South Carolina	Piedmont	82°58'43" W	34°49'26" N	17	2	1	3.96	112	32	16
14	S13	Georgia	Blue Ridge	83°20'19" W	34°26'47" N	18	0	1	18.63	61	9	3
15	S130	South Carolina	Blue Ridge	82°58'01" W	34°32'37" N	19	2	1	15.93	54	16	1
16	S131	South Carolina	Piedmont	82°57'22" W	34°32'45" N	20	2	2	4.59	62	22	8
17	S132	South Carolina	Piedmont	82°57'23" W	34°31'29" N	21	2	2	7.88	48	19	8
18	S135	Georgia	Piedmont	82°58'59" W	33°46'26" N	23	2	2	2.79	75	22	7
19	S136	South Carolina	Piedmont	82°46'53" W	34°54'04" N	24	2	1	2.03	102	19	11
20	S138	South Carolina	Piedmont	82°45'39" W	34°55'13" N	25	2	2	5.8	82	26	16

The table contains the following:

- A column called Station ID that uniquely identifies each sample.
- A column group called Station Descriptors consisting of seven columns that describe aspects of where a sample was taken. (To make such a group, select multiple columns in the **Columns** panel by clicking and holding down the Shift key, and then right-click and select **Group Columns** from the context-sensitive menu. Once the group has been made, you can double-click on the assigned name to enter a more descriptive name.)

- A group of columns called **Ys** consisting of four columns that describe the water quality of each sample. (See Figure 7.2.)
- A group of columns called **Xs** consisting of 26 columns that describe the landscape conditions associated with each sample. (See Figure 7.3.)

In total, the table WaterQuality1.jmp contains 38 columns. Brief descriptions of the Ys and Xs are provided in Figures 7.2 and 7.3. More specific descriptions of the variables as well as references relating to the specific measurement protocols can be found in Nash and Chaloud (2011). Brief descriptions of the variables are entered as **Notes** under **Column Info** for the relevant columns in the data table.

Figure 7.2: Description of Ys

Column Name	Full Name	Description
AGPT	Algal Growth Potential Test	An indicator of the level of nutrients that are biologically-available to support algal growth. Higher levels of nutrients are indicated by higher values.
HAB	Macroinvertebrate Habitat	A weighted composite score derived from visual observations of stream habitat characteristics. Higher scores indicate better habitat conditions for macroinvertebrate populations.
RICH	Macroinvertebrate Species Richness	A count of the number of taxa observed in a sample collected from a 100 meter stream segment. Higher numbers indicate greater diversity. For this study, counts exceeding 26 indicated non-impaired conditions, while counts below 11 indicated severely impaired conditions.
EPT	Ephemeroptera/Plecoptera/Trichoptera Index	An index derived by assessing the density of three orders of macroinvertebrate that are known to be sensitive to environmental conditions. The orders are: Ephemeroptera (mayflies), Plecoptera (stoneflies), and Trichoptera (caddisflies). For this study, values exceeding 10% indicate non-impaired conditions, while values of 1% or below indicate severely impaired conditions.

Figure 7.3: Description of Xs

Column Name	Brief Description
c	Percent crop
p	Percent pasture
b	Percent barren
u	Percent urban
f	Percent forest
q	Percent wetlands
w	Percent water
ah	Agriculture on highly erodible soils
az	Agriculture on slopes >3%
azh	Agriculture on slopes >3% with highly erodible soils
am	Agriculture on moderately erodible soils
azm	Agriculture on slopes >3% with moderately erodible soils
bzh	Barren on slopes >3% and highly erodible soils
bzm	Barren on slopes >3% with moderately erodible soils
cz	Crops on slopes >3%
czm	Crops on slopes >3% with moderately erodible soils
pz	Pasture on slopes >3%
e	Erodible soils
z	Slope >3%
x	Mean slope
s	Standard deviation slope
zm	Moderately erodible soils on slopes >3%
d	Stream density
v	Total road length within 30 meters of streams
r	Total road length in watershed
t	Total power, pipe, and telephone line length in watershed

You will note that 11 columns in the group of Xs have names that consist of two or three characters. These composite-character variables are variations on the base (single-character) variables. For example, the variable z represents the percent of the total area with slope exceeding 3%. The variable az represents the percent of the total area on slopes exceeding 3% that is used for agriculture. Because of the nature of these composite variables, there will be correlation with the underlying variables involved in their definitions.

If you run **Distribution** on the column Ecoregion, found in the Station Descriptors group, you see that three regions are represented: "Blue Ridge", "Piedmont", and "Coastal Region". The rows in WaterQuality1.jmp have been colored by the column Ecoregion. (To color the rows yourself, select **Rows > Color or Mark by Column**.) In addition, a **Value Colors** property has been assigned to Ecoregion to make the colors that JMP uses more interpretable. By clicking on the asterisk next to Ecoregion, found in the Station Descriptors group in the **Columns** panel, and selecting **Value Colors**, you see that compatible colors were assigned to the regions (Figure 7.4).

Figure 7.4: Value Colors for Ecoregion

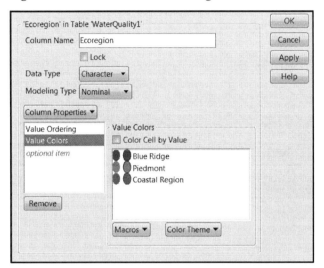

Initial Data Visualization

Location of Field Stations

The Savannah River is located in the southeastern United States, where it forms most of the border between the states of Georgia and South Carolina. The two large cities of Augusta and Savannah, Georgia, are located on the river. Within the Savannah River Basin, there are three distinct spatial patterns distinguished by the amount of forest cover and wetlands. The three regions can be labeled Blue Ridge, Piedmont, and Coastal Plain.

In the Blue Ridge Mountains, the home of the Savannah River headwaters, evergreen forests predominate. This landscape gives way to the Piedmont, an area where pastureland and hay fields, mixed and deciduous forests, as well as parks and recreational areas and several urban areas are found. There are two large reservoirs on the main river. South of Augusta, the Coastal Plain begins, evidenced by crop agriculture and wetlands. The mouth of the river broadens to an estuary southeast of Savannah.

Below Augusta, Georgia, extensive row crop agriculture is evident, along with wetland areas. The city of Savannah is located near the outlet of the river to the Atlantic Ocean. The spatial patterns seen in the landcover define three ecoregions: Blue Ridge, Piedmont, and Coastal Plain.

To see the geographical distribution of the sampling locations, run the first saved script, **Field Stations**, to produce Figure 7.5. You can see that the 86 locations are spread over the two states that make up the Savannah River basin, Georgia and South Carolina, and the three ecoregions.

Figure 7.5: Location of Field Stations within State and Ecoregion

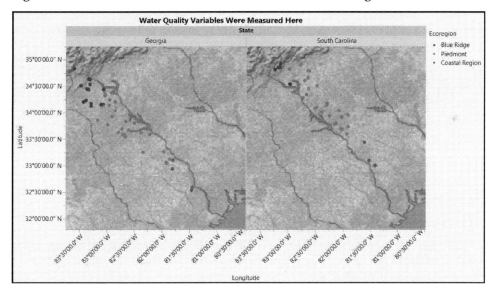

We expect that Ecoregion might have an impact on modeling, but that State will not.

Missing Response Values

Next we investigate the pattern of missing data.

1. Select **Tables > Missing Data Pattern.**
2. Select the column groups Ys and Xs, and click **Add Columns**.
3. Click **OK**.

(Alternatively, run the saved script Missing Data.) The data table shown in Figure 7.6 appears. Rows 2 and 3 of this data table indicate that only six of the 86 samples have one or two missing values for the water quality variables (Ys). There are no Xs missing.

Figure 7.6: Six Samples Have Missing Values for Water Quality Variables

- Missing Data Pattern
- Source
- Treemap
- Cell Plot

	Count	Number of columns missing	Patterns	AGPT	HAB	RICH	EPT	c	p
1	80	0	000000000000000000000000000000	0	0	0	0	0	0
2	1	2	001100000000000000000000000000	0	0	1	1	0	0
3	5	1	010000000000000000000000000000	0	1	0	0	0	0

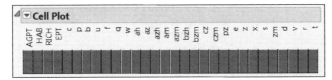

Run the Cell Plot script in the Missing Data Pattern table for a visual indication of where values are missing. It is easy to see now that five rows have missing values only for HAB, while one row has missing values for both RICH and EPT. Note that it is typical to have missing values when data are collected in the field.

Figure 7.7: Cell Plot for Missing Data Pattern Table

Impute Missing Data

As with any multivariate analysis, appropriate handling of missing values is an important issue. The PLS platform in JMP Pro provides two imputation methods: Mean and EM. These can be accessed from both **Analyze > Multivariate Methods > Partial Least Squares** and **Analyze > Fit Model**.

To impute missing data, check the **Impute Missing Data** check box on the launch window. If this box is not checked, rows that contain missing values for either **X**s or **Y**s are not included in the analysis. When you check **Impute Missing Data**, you are asked to select which **Imputation Method** to use:

- **Mean**: This option replaces the missing value in a column with the mean of the nonmissing values in the column.
- **EM**: This option imputes missing values using an iterative Expectation-Maximizaton (EM) method. On the first iteration, the model is fit to the data with missing values replaced by their column means. Missing values are imputed using the predictions from this model. On following iterations, the predictions

from the previous model fit are used to obtain new predicted values. When the iteration stops, the predicted values become the imputed values.

Note that the EM method depends on the type of fit (NIPALS or SIMPLS) selected and the number of factors specified. When you select the EM method, you can set the number of iterations. When you run your analysis, a **Missing Value Imputation** report appears. This report shows you the maximum difference between missing value predictions at each stage of the iteration. The iterations are terminated when that maximum difference for both **X**s and **Y**s falls below 0.00000001.

In this chapter we will assume that you are using JMP Pro. If you are using JMP, the six rows identified in Figure 7.6 will be ignored.

Distributions

Continuing to visualize the data, let's look at the distributions of the variables in the Ys group and Xs group. Complete the following steps, or run the saved script Univariate Distributions.

1. Select **Analyze > Distribution**.
2. At the lower left of the window, select the **Histograms Only** check box.
3. Select the two column groups called Ys and Xs and click **Y, Columns**.
4. Click **OK**.
5. You will obtain the plots partially shown in Figure 7.8.

Figure 7.8: Distributions of Responses and Predictors (Partial View)

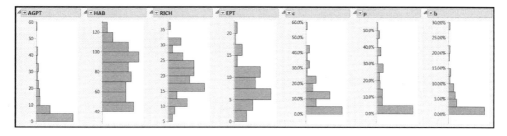

With the exception of AGPT, the distributions of the Ys appear fairly mound-shaped. Because our analysis is more reliable when the responses have at least approximately a multivariate normal distribution, we will apply a transformation to AGPT to bring it closer to normality.

Although many of the Xs also have distributions that are skewed and unruly, we find this behavior less troubling than unruly distributions for the Ys. However, note that most of the landscape variables assume values of zero, and some have a high percentage of zeros. To verify this, hold down the Ctrl key while clicking the red triangle corresponding to any one of the variables. Then select **Display Options > Quantiles**. These substantial percentages of zeros make it very difficult to transform the landscape variables to mound-shaped distributions.

Transforming AGPT

To make AGPT more symmetric, we apply a logarithmic transformation. If you are using JMP 11, you can create transformation columns in launch windows. We illustrate this approach first. Then we show the more general approach of defining the transformation using a column formula. (If you want to bypass this discussion, you can run the script Add Log[AGPT]).

Transforming through a Launch Window

1. Select **Analyze > Distribution**.

2. Click the disclosure icon next to Ys to reveal the column names.

3. Right-click AGPT and select **Transform > Log**. This selection creates a virtual column called Log[AGPT]. You can use this column in your analysis, but it is not saved to the data table.

4. To save the virtual column to your data table, right-click on Log[AGPT] and select **Add to Data Table**.

5. Add Log[AGPT] to the **Y, Columns** list.

6. Click **OK**.

You have created and saved the new column, Log[AGPT], and you have produced a plot of its distribution (Figure 7.10).

Transforming by Creating a Column Formula

1. Double-click in the column heading area to the right of the last column in the table to insert a new column called Column 39.

2. Right-click in the column heading area and select **Formula** from the pop-up menu to open the formula editor window.

3. In the formula editor, select AGPT from the list of **Table Columns**. This selection inserts AGPT into a red box in the formula editing area. The red box indicates which formula element is selected.

4. From the list of **Functions**, select **Transcendental > Log**. This selection applies the natural log transformation to AGPT, as shown in the formula area.

5. Click **OK**.

6. Double-click on the column heading to rename the column. We will call it Log[AGPT].

In the **Columns** panel in the data table, the new column Log[AGPT] appears with an icon resembling a large plus sign to its right. This icon indicates that the column is defined by a formula. Clicking the + sign selects the column and displays the formula (Figure 7.9).

Figure 7.9: Plus Sign Indicating Formula for Log[AGPT]

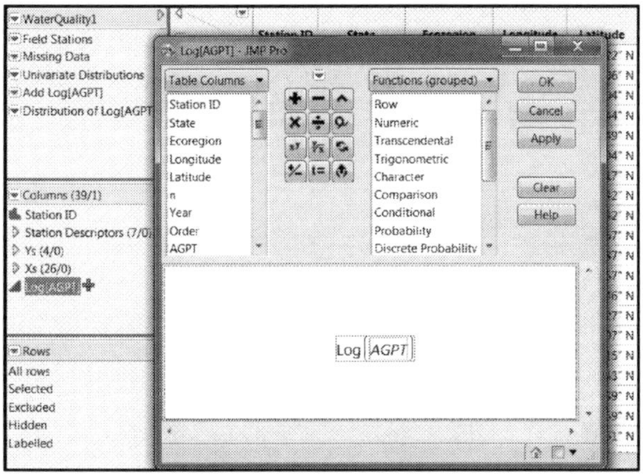

Suppose that you want to move this column, say to place it closer to the original four responses. In the **Columns** panel, click and drag it to place it where you would like it.

To check the effect of the transformation on the distribution of AGPT, select **Analyze > Distribution** and enter Log[AGPT] in the **Y, Columns** box. Note that, if you transformed AGPT in the Distribution launch window, you have already produced this plot. (Alternatively, run the script Distribution of Log[AGPT].) The distribution, shown in Figure 7.10, indicates that the right-skewed distribution of AGPT has been transformed to a more mound-shaped distribution.

Figure 7.10: Distribution of Log[AGPT]

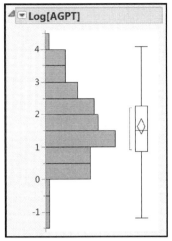

Keep in mind that we have transformed AGPT for modeling purposes. It is important to remember that you need to apply the inverse transformation to Log[AGPT] when interpreting analytical results in terms of the raw measurements.

Differences by Ecoregion

At this point, close the data table WaterQuality1.jmp and click on the master journal link for WaterQuality2.jmp. Here, AGPT has been replaced by its transformed analog, Log(AGPT), in the column group called Ys. (Note that we are using the more conventional functional notation, Log(AGPT), rather than the square bracket notation.)

Next we turn attention to looking for any differences in water quality among the three ecoregions.

1. Select **Analyze > Fit Y by X.**
2. Select the Ys group from the **Select Columns** list and click **Y, Response**.
3. Select Ecoregion from the Station Descriptors group in the **Select Columns** list and click **X, Factor**.
4. Click **OK**.
5. While holding down the Ctrl key, select **Densities > Compare Densities** from any of the **Oneway** red triangle menus.

Holding down the Ctrl key while selecting an option *broadcasts* this option to all relevant reports. Check to see that a **Compare Densities** plot has been added for each response.

Alternatively, run the saved script Distribution of Ys by Ecoregion. The report appears as shown in Figure 7.11.

Figure 7.11: Distribution of Responses within and across Ecoregions

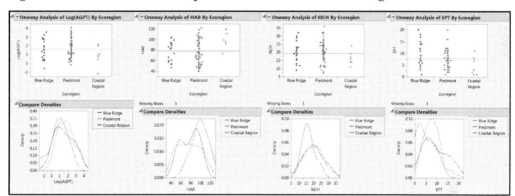

Note that, with the possible exception of HAB and perhaps EPT, the responses show little variation across Ecoregion. HAB seems to have higher values in the Coastal Region than do the other three responses, while EPT might have lower values in the Coastal Region than do the other responses.

Following the same procedure for the Xs group as we did for the Ys, or by running the saved script Distribution of Xs by Ecoregion, obtain the report partially shown in Figure 7.12.

Figure 7.12: Distribution of Predictors within and across Ecoregions (Partial View)

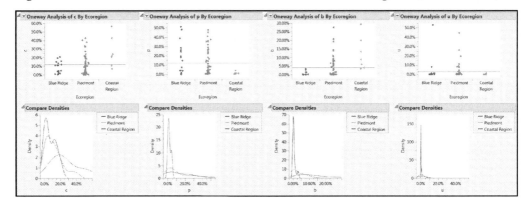

The plots show that the Piedmont region had the largest number of water samples and the Coastal Region the least, information that we have already seen in Figure 7.5. But we also see that, as should be expected, the variation of the landscape variables across the three ecoregions is considerable.

Visualization of Two Variables at a Time

Now we explore the relationships between pairs of variables. We construct two scatterplot matrices, one for the Ys and one for the Xs. To construct the plot for the Ys, complete the following steps:

1. Select **Graph > Scatterplot Matrix**.
2. Select the Ys group from the **Select Columns** list.
3. Click **Y, Columns**.
4. From the **Matrix Format** menu in the lower left of the window, select **Square**.
5. Click **OK**.
6. From the report's red triangle menu, select **Group By**.

7. In the window that opens, select **Grouped by Column** and select Ecoregion. Note that **Coverage** is set, by default, to 0.95.

8. Click **OK**.

9. From the **Scatterplot Matrix** report's red triangle menu, select **Density Ellipses > Density Ellipses** and **Density Ellipses > Shaded Ellipses**.

The report shown in Figure 7.13 appears. Constructing the scatterplot matrix for the Xs is done in a similar fashion (partially shown in Figure 7.14). Alternatively, run the scripts Scatterplot Matrix for Ys and Scatterplot Matrix for Xs.

Figure 7.13: Pairwise Variation of Responses within and across Ecoregions

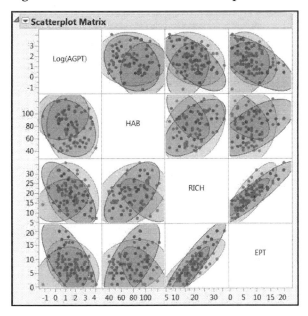

Figure 7.14: Pairwise Variation of Predictors within and across Ecoregions (Partial View)

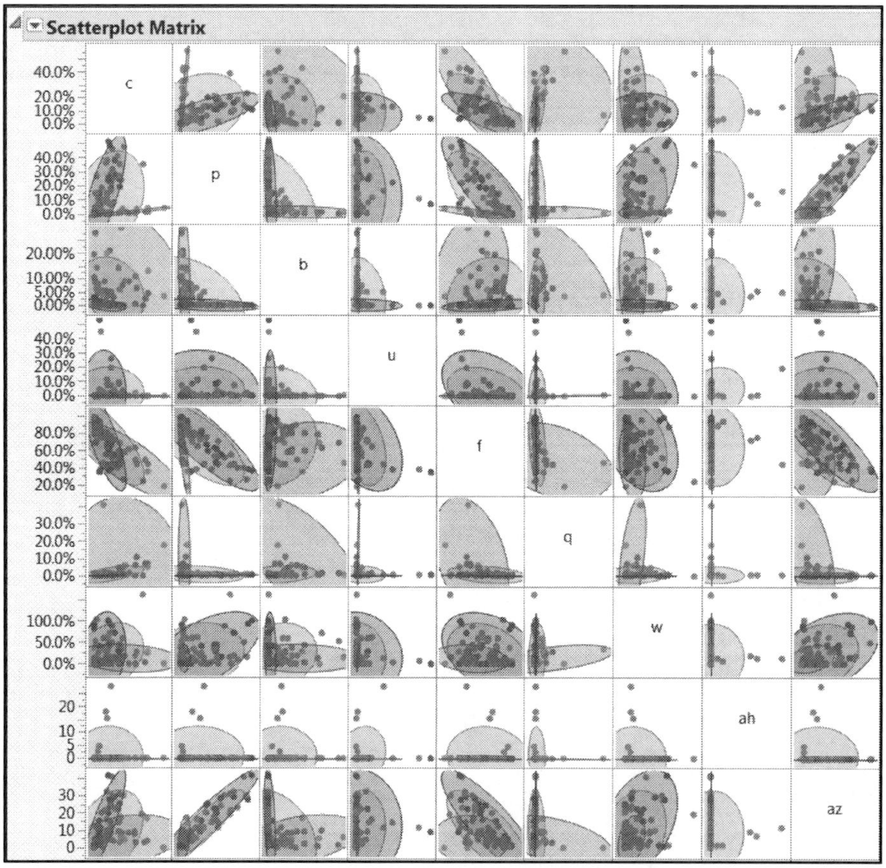

The density ellipses shown in Figures 7.13 and 7.14 delineate 95% confidence regions for the joint distribution of points, assuming that these points come from a bivariate normal distribution. We are somewhat comfortable with this interpretation for the Ys, because these are generally at least mound-shaped. However, the ellipses have little statistical relevance for the Xs, because their distributions are generally far from normal. For the Xs, these ellipses serve only as a useful guide to the eye. You can see that, broadly, the responses lie in somewhat similar regions, but that in many cases the predictors show marked differences by Ecoregion.

Conclusions from Visual Analysis and Implications

This visualization of the data alerts us to the fact that Ecoregion has an important impact on the patterns of variation in the data. This impact suggests that we consider two modeling approaches:

1. Produce a model for the Savannah River basin as a whole, but include Ecoregion as an additional **X**.

2. Produce three models, one for each ecoregion.

Approach (2) necessarily restricts the number of observations available for building and testing each model to the number of observations made on each ecoregion. But, even so, it might provide better predictions than approach (1) for similar ecoregions located elsewhere. Approach (1) gives us more data to work with and might produce an omnibus model that has more practical utility, assuming it can predict well.

Following Nash and Chaloud (2011), we will explore both approaches. However, in the interest of brevity, we will restrict approach (2) to the study of a single ecoregion, namely the Blue Ridge.

A First PLS Model for the Savannah River Basin

Our Plan

We will begin by developing a PLS model based on data from the entire Savannah River Basin using approach (1).

The last column in WaterQuality2.jmp is called Test Set. It contains the values "0" (labeled as "No") and "1" (labeled as "Yes"). This column was constructed by selecting a stratified random sample of the full table, taking a proportion of 0.3 of the rows for each level of Ecoregion. You can review the outcome of this random selection by selecting **Analyze > Distribution** to look at the distributions of Test Set and Ecoregion. Clicking on the **Yes** bar shows that there is an appropriate number of rows highlighted for each level of Ecoregion.

If you want to construct the Test Set column yourself, complete the following steps:

1. Select **Tables > Subset**.

2. Click the button next to **Random - sampling rate**. Specify a random sampling rate of **0.3**.

3. Select the **Stratify** check box and then select the Ecoregion column.

4. Select the **Link to Original Table** check box and click **OK** to make a new subset table.

5. In the subset table, select **Rows > Row Selection > Select All Rows**. Note that, because the tables are linked, this selects the corresponding rows in WaterQuality2.jmp as well.

6. Close the subset table, then select **Rows > Row Selection > Name Selection in Column**. Call the new column Test Set 2, and click **OK**. You can also add a **Value Label** property to Test Set 2 to display "Yes" and "No" rather than "1" and "0".

As in the Spearheads.jmp example in Chapter 1, we will use the Test Set column to provide development and test data for the modeling process. The data used for model building (training and validation) will consist of those rows for which Test Set has the label "No". The test set will consist of rows for which Test Set has the label "Yes" and will be used to test the predictive ability of the resulting model. (Note that the data table contains the saved scripts Use Only Non-Test Data, Use Only Test Data, and Use all Data. These scripts manipulate the **Exclude** and **Hide** row state attributes, as their names imply.)

We will emulate a situation where the data used for model building is the currently available data, while the test data only becomes available later on. To this end, WaterQuality2_Train.jmp contains the data we will use for modeling, and WaterQuality2_Test.jmp contains the test data.

Performing the Analysis

At this point, close WaterQuality2.jmp and open the data table WaterQuality2_Train.jmp. This data table contains only the 61 rows designated as not belonging to the test set in WaterQuality2.jmp.

In our model, we will be including a categorical variable, Ecoregion, as a predictor. Although JMP Pro supports categorical predictors, JMP does not. If you are using JMP rather than JMP Pro, skip ahead to the analysis in the section "A First PLS Model for the Blue Ridge Ecoregion." Or, follow along without using Ecoregion as a predictor. You can easily adapt the following steps using the PLS platform by selecting **Analyze > Multivariate Methods > Partial Least Squares**. Follow these steps if you have JMP Pro:

1. Select **Analyze > Fit Model**.
2. Enter the group called Ys as **Y**.
3. Enter Ecoregion, found in the Station Descriptors column group, and the column group Xs in the **Construct Model Effects** box.
4. Select **Partial Least Squares** as the **Personality**.
5. Select the **Impute Missing Data** check box.
6. Select **EM** as the **Imputation Method**.
7. Enter **20** for **Max Iterations**.
   Your window should appear as shown in Figure 7.15.
8. Click **Run**.

You can also run the script Fit Model Launch to obtain the Fit Model launch window.

Figure 7.15: Fit Model Window

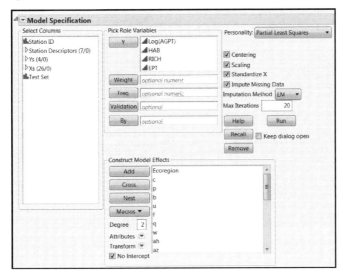

158 *Discovering Partial Least Squares with JMP*

The Model Launch control panel opens (Figure 7.16). Accept the default settings shown here. However, to obtain exactly the output shown below, complete the following steps:

1. Select the **Set Random Seed** option from the Partial Least Squares red triangle menu.
2. Enter the value **666**.
3. Click **OK**.
4. Click **Go**.

Figure 7.16: PLS Model Launch Control Panel

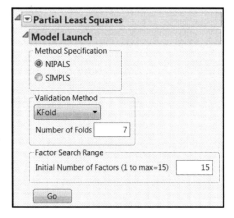

The Partial Least Squares Report

Clicking **Go** adds three report sections. (The saved script is PLS Fit.)

- A **Model Comparison Summary** report, which is updated when new fits are performed.
- A report with the results of your cross validation study. Because you selected the NIPALS method and accepted the default number of folds, namely, 7, this report is called **KFold Cross Validation with K=7 and Method=NIPALS**.
- A report giving results for a model with the number of factors chosen through the cross validation procedure. In this example, this report is entitled **NIPALS Fit with 1 Factors**, because you requested a NIPALS fit and the number of factors chosen by the cross validation procedure is one. Note that only the title bar for this last report is shown in Figure 7.17.

Figure 7.17: PLS Report for One-Factor Fit

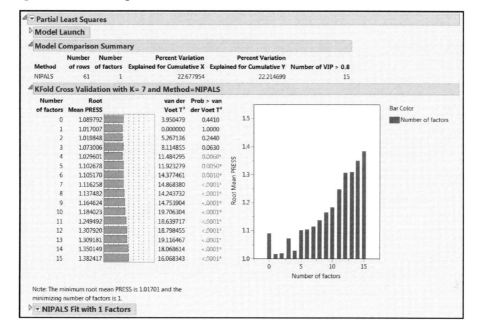

The **Model Comparison Summary** gives details for the automatically chosen fit, and shows that the single factor accounts for about 23% of the variation in the predictors and about 22% of the variation in the responses. The last column, **Number of VIP > 0.8**, indicates that 15 of the 27 predictors are influential in determining the factor. There might be an opportunity for refining the model by dropping some of the predictors.

The NIPALS Fit Report

Now let's look at the report for the fit, **NIPALS Fit with 1 Factors** (Figure 7.18). The first report is the **Missing Value Imputation** report. This report gives information about how the behavior of the imputation changed over the 20 iterations that we specified. Notice that only 11 iterations are listed. This is because the convergence criterion of a maximum absolute difference for imputed values of .00000001 was achieved after the 11th iteration.

The **X-Y Scores Plots** report shows clear correlation between the X and Y scores.

The **Percent Variation Explained** report displays the percent of variation in the Xs and Ys that is explained by the single factor. The **Cumulative X** and **Cumulative Y** values agree with the corresponding figures in the **Model Comparison Summary**.

The two **Model Coefficients** reports give the estimated model coefficients for predicting the **Y**s. **The Model Coefficients for Centered and Scaled Data** report shows the coefficients that apply when both the **X**s and **Y**s have been standardized to have mean zero and standard deviation one. The **Model Coefficients for Original Data** report expresses the model in terms of the standardized **X** values, because **Standardize X** was selected on the Fit Model launch window. This second set of model coefficient estimates is often of secondary interest.

Figure 7.18: NIPALS Fit Report

Iteration	Max Absolute Imputation Difference for X Effects	Max Absolute Imputation Difference for Y Responses	Percent of Nonmissing Variation Explained for X Effects	Percent of Nonmissing Variation Explained for Y Responses
1.0000	0.0000	0.6026	44.1561	26.9310
2.0000	0.0000	0.0641	44.1233	27.0980
3.0000	0.0000	0.0062	44.1255	27.1263
4.0000	0.0000	0.0006	44.1262	27.1287
5.0000	0.0000	0.5806	22.7115	22.3822
6.0000	0.0000	0.0170	22.6798	22.2150
7.0000	0.0000	0.0009	22.6781	22.2147
8.0000	0.0000	0.0001	22.6780	22.2147
9.0000	0.0000	0.0000	22.6780	22.2147
10.0000	0.0000	0.0000	22.6780	22.2147
11.0000	0.0000	0.0000	22.6780	22.2147

To get a better sense of how the data relate to the one factor model, we look at the **T Square Plot** and the **Diagnostics Plots**. Select each of these in turn from the red triangle menu for the **NIPALS Fit with 1 Factors** report. (Or, run the script Diagnostic Plots.) The T Square Plot is shown in Figure 7.19.

Figure 7.19: T Square Plot

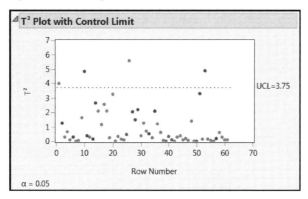

The T Square plot shows a multivariate distance for the X score corresponding to each observation. There is one point that is just beyond the upper limit and three that are a little further beyond.

The upper limit (**UCL**) on the T Square plot is computed assuming that the X scores have approximately a multivariate normal distribution. In this model, there is only one factor, so this amounts to assuming that the scores have a normal distribution. To verify that the X scores are at least mound-shaped, select **Save Columns > Save Scores** from the red triangle menu. This selection saves the X and Y scores to the data table. Then select **Analyze > Distribution** to obtain a histogram of X Scores 1.

Drag a rectangle to select the four points that fall above the upper limit in the T Square plot. In the Diagnostics Plots, all four points fall at the extremes for the distributions of the predicted responses (Figure 7.20). Even so, based on the Diagnostics Plots, none of the four points seem unusual or influential enough to cause concern.

Figure 7.20: Diagnostics Plots with Four Rows Selected

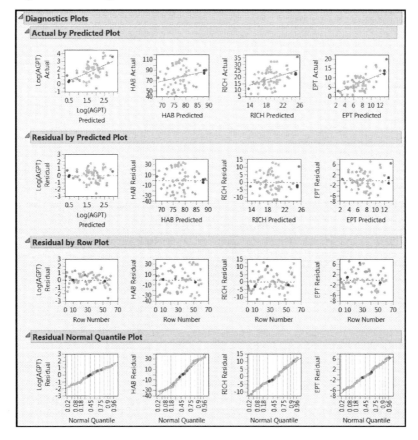

However, the **Actual by Predicted Plot** portion of the **Diagnostics Plots** report is of interest. The red lines in the plots are at the diagonal. If the responses were being perfectly predicted, the points would fall on these lines. Note that predicted values for HAB and RICH appear to be biased. Predictions for small values are larger than expected, and predictions for large values are smaller than expected.

This bias could be connected with the number of factors used or the fact that all four responses are being jointly modeled. In Appendix 2, we discuss the bias-variance tradeoff in greater detail. There we present a script, ComparePLS1andPLS2, that you can use with your own data to explore the impact of the number of factors and joint modeling of multiple responses.

Yet another way to check for outliers is to look at the Euclidean distance from each observation to the PLS model in both the **X** and **Y** spaces. No point should be dramatically farther from the model than the rest. A point that is unusually distant might be unduly influencing the fit of the model. Or, if there is a cluster of distant points, it might be that they have something in common and should be analyzed separately. These distances are plotted in the Distance Plots.

Select the **Distance Plots** option from the **NIPALS Fit with 1 Factors** red triangle menu. Again, none of the selected points seem unusual enough to cause concern. Note that these distances to the model are called *DModX* and *DModY* by Umetrics and others (Eriksson et al. 2006).

Figure 7.21: Distance Plots with Four Rows Selected

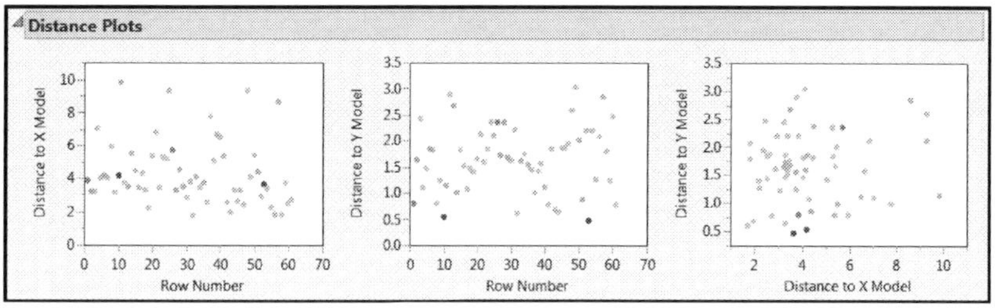

Defining a Pruned Model

To explore the VIPs for the predictors, select the **VIP vs Coefficients Plots** option from the **NIPALS Fit with 1 Factors** red triangle menu (Figure 7.22). (Alternatively, run the script VIP Plots.) Consider the plot in the **VIP vs Coefficients for Centered and Scaled Data** report.

Figure 7.22: VIP vs Coefficients Plot for Centered and Scaled Data

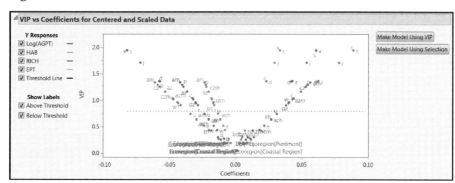

The **Y Responses** check boxes enable you to review each response in isolation. In this case, all four responses show a characteristic "V" shape, indicating that the terms that contribute to the regression (the terms with larger absolute coefficients) also are important relative to dimensionality reduction in PLS. Note that labels are not shown for points that are very close together. However, tooltips enable you to identify which model term corresponds to any given point in the plot.

Contrary to our expectation, the three terms associated with Ecoregion all have small coefficients and VIP values. We suspect that this is because other continuous predictors differentiate the varied landscape conditions found in the three ecoregions.

We note that s, x, f, z, and am have large VIPs for most responses. The predictors s, x, and z all deal with slope. They have negative coefficients for Log(AGPT), which measures habitat conditions, but positive coefficients for the other three responses. The predictor f, which measures percent forest, behaves similarly, as one might expect. However, am, which measures agriculture on moderately erodible soils, has a positive coefficient for Log(AGPT) and negative coefficients for the remaining three responses. The **VIP vs Coefficients** plot shows a similar differentiation for other predictors with high VIPs that deal with crops and agriculture (cz, czm, az, azm, and c). Information of this type should be very meaningful to an ecologist.

The two buttons to the right of the plot in Figure 7.22 are convenient for producing a pruned model. We will use an alternative VIP threshold of 1.0, which is sometimes recommended in the PLS literature, rather than the JMP default value of 0.8. (For a comparison of the 0.8 and 1.0 cut-offs, see Appendix 1.) In part, we use the 1.0 cut-off to illustrate selecting a specific set of predictors.

To select those terms that have VIP values of 1.0 or larger, use your cursor to drag a rectangle around the desired points. (See Figure 7.23.) Then, with these terms selected, click **Make Model Using Selection** to the right of the plot. We have selected 12 terms in this fashion. (The script is Pruned Fit Model Launch.)

Figure 7.23: Selecting Terms for the Pruned Model

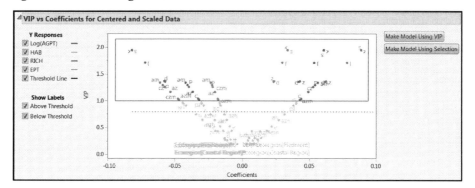

Dragging a rectangle to enclose the desired points might not be sufficiently precise, especially when many terms are present. We describe a method that offers more control, should it be needed:

1. Select the **Variable Importance Plot** option from the red triangle menu for the relevant model fit.

2. Below the **Variable Importance Plot**, there is a **Variable Importance Table** report. Open the report. This table gives the VIPs for each term.

3. Right-click in this table and select **Make Into Data Table** from the context-sensitive menu.

4. Right-click the VIP column and select **Sort > Descending**.

5. Select the rows with the desired VIP values (in our example, rows 1 to 12).

6. Select column X so that the terms of interest are selected.

7. Right-click in the selected area of the table body and select **Copy** from the context-sensitive menu.

8. Close this auxiliary table.

9. Making sure that your working data table is active, select **Analyze > Fit Model**.

10. Right-click in the white region of the **Construct Model Effects** box and select **Paste**.

11. Complete the Fit Model window as required.

When you complete these steps for the one-factor model using a VIP cut-off of 1.0, you select a total of 12 terms. These terms and the completed Fit Model window are shown in Figure 2.24. Remember to set the **Imputation Method** to **EM** and to enter **20** as the number of **Max Iterations**.

Figure 7.24: Fit Model Window for Pruned Model

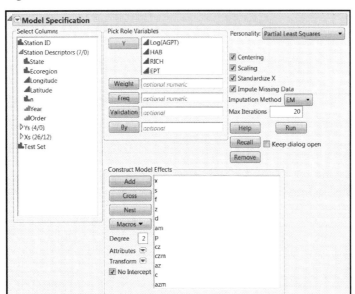

A Pruned PLS Model for the Savannah River Basin

Model Fit

To fit the pruned model with the chosen terms, click **Run** in the Fit Model launch window that results from clicking the **Make Model Using Selection** button as described earlier, from your constructed Fit Model window, or from running the script Pruned Fit Model Launch. From the red triangle menu in the report, select **Set Random Seed** and enter the random seed **666** as described earlier to replicate the results shown here.

In the PLS Model Launch control panel, accept the defaults and click **Go**. This produces the PLS Report shown in Figure 7.25. You can also generate this report by using the Pruned Fit script in the data table.

Figure 7.25: PLS Report for Pruned Model

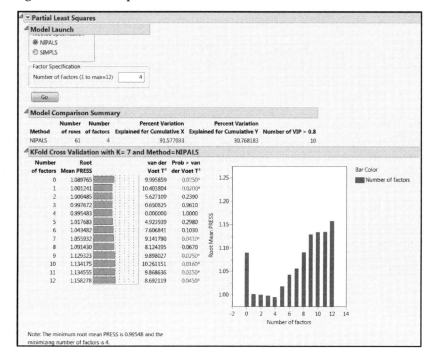

The cross validation procedure selects a four-factor model, which explains almost 92% of the variation in the **X**s and almost 31% of the variation in the **Y**s. For comparison, note that Table 4 of Nash and Chaloud (2011) states that 80% and 43% of the variation in the **X**s and **Y**s is explained by their model for the whole Savannah River basin. But note also that their model restricts attention to only two of the four **Y**s (HAB and EPT).

This observation raises the question of whether we can produce better models by using the PLS framework to model each **Y** separately, rather than modeling all the **Y**s collectively. For example, a three-factor model for EPT as a single **Y** accounts for 62% of the variation in **Y**. We will return to this interesting question in Appendix 2, where we present a script that you can use to compare the results of fitting each **Y** with a separate model to the results of fitting all **Y**s together in the same model.

The van der Voet test in the **KFold Cross Validation** report indicates that the four-factor model Root Mean PRESS does not differ significantly from that of the two-factor model. In the interests of parsimony, we will adopt a two-factor model. To fit a two-factor model:

168 *Discovering Partial Least Squares with JMP*

1. In the **Factor Specification** area of the Model Launch control panel, enter **2** as the **Number of Factors**.
2. Click **Go**.

Alternatively, run the saved script Pruned Fit: Two Factors.

Diagnostics

From the red triangle menu next to the **NIPALS Fit with 2 Factors** report, select **Diagnostics Plots** to produce the report shown in Figure 7.26. The **Residual by Predicted Plot** report shows no anomalies or patterns. You can check that the **T Square Plot** and the **Distance Plots** show no egregious or problematic points.

However, the **Actual by Predicted Plot** report once again indicates that HAB and RICH are not well predicted by this model. In fact, predictions for all four responses seem somewhat biased: higher response values are given lower predictions, and lower response values are given higher predictions. If interest is in predicting the individual measures, it might be better to model these responses separately. We encourage you to explore various options using the script ComparePLS1andPLS2, described in Appendix 2.

Figure 7.26: Diagnostic Plots for Two-Factor Pruned Model

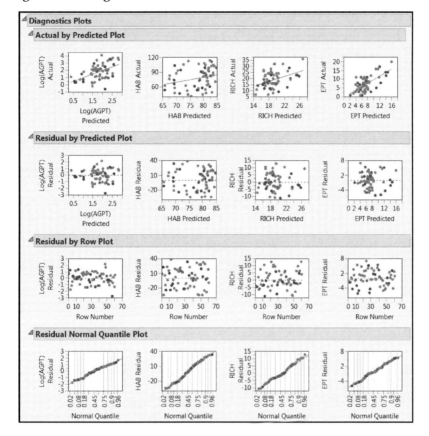

Saving the Prediction Formulas

In the next section, we will validate our model using the test set data. To this end, we must save the prediction formulas for the **Y**s. To do this, you can follow the instructions below or run the script Add Predictions for PLS Pruned Model.

From the red triangle menu next to **NIPALS Fit with 2 Factors**, select **Save Columns > Save Prediction Formula**. This script adds four columns, one for each response to the end of the data table WaterQuality2_Train.jmp. These columns are formula columns. You can view the formulas by clicking the **+** sign next to each column name in the **Columns** panel.

Next, place these four prediction columns in a column group. In the **Columns** panel, select the four columns. Right-click on the selected column names, and select **Group**

Columns from the menu. This selection groups the columns under the name Pred Formula Log(AGPT) etc. Click on this name, and rename the group to Predictions from Pruned Model.

Comparing Actual Values to Predicted Values for the Test Set

At this point, open the test set that we have held in reserve, WaterQuality2_Test.jmp. You will use this independent data to test the model's performance on new data.

To apply the formulas saved in WaterQuality2_Train.jmp to the 25 rows in WaterQuality2_Test.jmp, you will append this second data table to WaterQuality2_Train.jmp. Complete the following steps, or run the script Add Test Data in WaterQuality2_Train.jmp once you have opened WaterQuality2_Test.jmp.

1. Ensure that both WaterQuality2_Train.jmp and WaterQuality2_Test.jmp are open.
2. Make WaterQuality2_Train.jmp the active table, and select **Tables > Concatenate**.
3. From the **Opened Data Table** list, select WaterQuality2_Test.jmp.
4. Click **Add**.
5. Check **Save and evaluate formulas**.
6. In the **Output table name** text area, enter WaterQuality3.jmp.

    The window should appear as shown in Figure 7.27.

7. Click **OK**.

Figure 7.27: Completed Concatenate Window

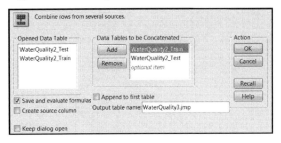

Check that the 25 rows have been appended to the new data table WaterQuality3.jmp, and that the prediction formulas have evaluated for these rows.

Chapter 7 ~ Water Quality in the Savannah River Basin **171**

By selecting **Graph > Graph Builder**, it is easy to produce plots for the test data of actual versus predicted values for each of the four responses. We will illustrate obtaining such a plot for Log(AGPT). To go directly to the plots, run the script **Actual by Predicted** in WaterQuality3.jmp.

1. Ensure that WaterQuality3.jmp is the active data table. Select **Graph > Graph Builder**.

2. From the **Variables** list that appears at the left, select Log(AGPT) and drag it to the **Y** area near the left vertical axis of the plot template. The various predicted values are plotted in a jittered fashion.

3. From the **Variables** list, select Pred Formula Log(AGPT) and drag it to the **X** area at the bottom of the plot template. Now the points are plotted as they would be in a scatterplot, except that a smoother has been added.

4. Click the Smoother to deselect it. The **Smoother** is the second icon from the left above the template.

   Next, we will plot a line on the diagonal. If the predicted values exactly matched the actual values, all points would fall on the diagonal.

5. Right-click in the plot and select **Customize**.

6. In the Customize Graph window, click the **+** sign to open a text window where you can enter a script.

7. In the text area, enter the following script, so that your window matches the one in Figure 7.28.
   Pen Color("Green");
   Y Function(x, x);

8. Click **OK**.

9. Resize the graph appropriately.

Figure 7.28: Customize Graph Window

The resulting plot displays results for all 81 rows. Run the script Use Only Test Data to display only the 25 test rows. The completed plot for Log(AGPT), along with those for the three other responses, is shown in Figure 7.29.

Figure 7.29: Actual by Predicted Plots for Test Set Using PLS Pruned Model

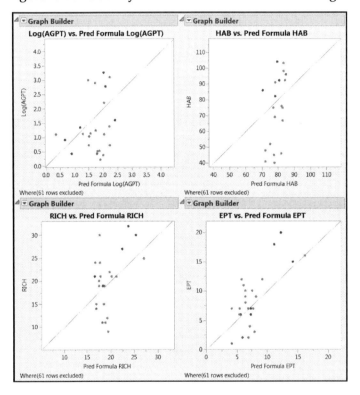

For all four responses, we see some bias. Specifically, higher values of the response are generally predicted to have lower values, and lower values are predicted to have higher values, relative to the line that describes perfect agreement. To see that this is consistent with the model for the training data, run the script Use Only Non-Test Data to exclude all but the 61 training rows. The **Actual by Predicted Plots** display is updated. The bias remains evident. We give further consideration to the question of bias in PLS predictions in Appendix 2.

A First PLS Model for the Blue Ridge Ecoregion

Making the Subset

At the end of the section "Conclusions from Visual Data Analysis and Implications," we noted two strategies for modeling. The first was to obtain a single model that covers all three ecoregions. The second was to obtain separate models for each ecoregion. In this section, we illustrate the second approach by building a model for one of the ecoregions, the Blue Ridge.

It is possible to build separate models for multiple ecoregions simultaneously from the same table. You can do this by assigning Ecoregion to the role of a **By** variable in the Fit Model window. However, this workflow leads to a proliferation of saved columns and the use of long names to distinguish them. For this reason, we will simply work with a subset of WaterQuality2.jmp that contains the appropriate rows.

To create the subset:

1. Open WaterQuality2.jmp by clicking on the link in the master journal.

    There are many ways to select the appropriate rows. We will use **Distribution**.

2. Select **Analyze > Distribution**.

3. Select Ecoregion from the column group Station Descriptors and click **Y, Columns**.

4. Click **OK**.

5. In the plot, double-click on the bar labeled "Blue Ridge".

Double-clicking produces the subset table shown in Figure 7.30.

Figure 7.30: Linked Subset Table Containing the Blue Ridge Data (Partial View)

	Station ID	State	Ecoregion	Longitude	Latitude	n
1	S09	Georgia	Blue Ridge	83°03'14" W	34°26'47" N	3
2	S11	Georgia	Blue Ridge	83°21'41" W	34°27'22" N	10
3	S121	Georgia	Blue Ridge	83°35'30" W	34°49'59" N	14
4	S122	Georgia	Blue Ridge	83°18'12" W	34°32'45" N	15
5	S123	Georgia	Blue Ridge	83°17'02" W	34°10'27" N	16
6	S13	Georgia	Blue Ridge	83°20'19" W	34°26'47" N	18
7	S130	South Carolina	Blue Ridge	82°58'01" W	34°32'37" N	19
8	S191	South Carolina	Blue Ridge	83°08'11" W	34°52'04" N	49
9	S192	South Carolina	Blue Ridge	83°09'59" W	34°51'41" N	50
10	S193	South Carolina	Blue Ridge	83°09'01" W	34°51'16" N	51
11	S196	South Carolina	Blue Ridge	83°14'11" W	34°48'55" N	54
12	S207	Georgia	Blue Ridge	83°22'44" W	34°14'14" N	58
13	S210	Georgia	Blue Ridge	83°03'59" W	34°09'47" N	60
14	S211	Georgia	Blue Ridge	83°06'05" W	34°09'24" N	61
15	S214	Georgia	Blue Ridge	83°17'19" W	34°07'49" N	63
16	S61	South Carolina	Blue Ridge	83°07'50" W	34°55'44" N	92
17	S68	Georgia	Blue Ridge	83°19'55" W	34°38'17" N	95
18	S69	Georgia	Blue Ridge	83°18'30" W	34°37'42" N	96
19	S72	Georgia	Blue Ridge	83°28'27" W	34°30'57" N	98
20	S83	Georgia	Blue Ridge	83°25'26" W	34°12'15" N	107

This table is a *linked* subset. If you close the parent table, the linked table closes too, so you need to save this table if you want to continue to work with it. Also, the scripts from WaterQuality2.jmp propagate to this new table. Some of these are no longer relevant to the reduced data and need to be removed.

At this point, we suggest that you close all open tables and open the table WaterQuality_BlueRidge.jmp by clicking on the link in the master journal. This table contains the data for the 20 Blue Ridge observations, and a set of relevant scripts. The column Test Set designates 14 observations that will be used for modeling and 6 for testing.

Reviewing the Data

The steps for reviewing the data and for building the PLS model mirror those we used for the data from the entire Savannah River Basin. For this reason, our instructions will be succinct and we will rely heavily on saved scripts.

Running the saved Missing Data script shows that two samples have missing values for HAB. We note that both of these are in the test set. Because coloring the rows by Ecoregion is no longer relevant, we choose to color the rows containing a missing value as red. To do this, in the **Missing Data Pattern** table produced by the Missing Data script, select **Rows > Color or Mark by Column**. Select the column Number of Columns Missing and click **OK**.

To review the data for the Blue Ridge ecoregion you can run the four scripts: Univariate Distribution of Ys, Univariate Distribution of Xs, Scatterplot Matrix for Ys, and Scatterplot Matrix for Xs. From the quantiles shown by the script Univariate Distribution of Xs, we see that the variables ah, azh, and bzh, which all deal with agriculture on highly erodible soils, have only zero values. As such, these are non-informative for modeling.

To remove these columns from further consideration, select all three columns in the **Columns** panel of the data table holding down the Ctrl key. Then, with your cursor on one of the highlighted selection areas, right-click and select **Exclude/Unexclude** from the context-sensitive menu. (Incidentally, if you include these columns in your model, JMP will ignore them. However, identifying such columns can be useful for your general understanding of the data.)

Performing the Analysis

First, we must hide and exclude the rows for which Test Set values are "No". To do this, select **Analyze > Distribution**, entering Test Set as **Y, Columns**, to select the "Yes" rows, and then **Hide and Exclude** them by right-clicking on one of the selection areas in the data table. Or, you can run the saved script Use Only Non-Test Data.

Populate the Fit Model launch window as shown in Figure 7.33, or run the saved script BR Fit Model Launch. Check to see that the three excluded columns do not appear in the group of Xs in the **Select Columns** list. You do not need to check the **Impute Missing Data** option, because there are no missing observations for the training data.

Figure 7.31: Fit Model Window

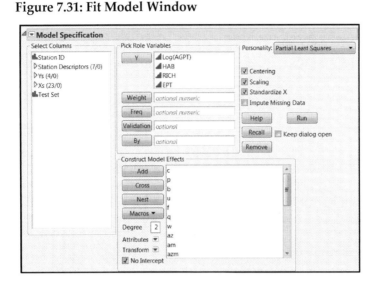

Click **Run** to obtain the Partial Least Squares Model Launch control panel. As before, to reproduce the results below exactly, set the random seed to be **666**. Click **Go** to accept the default settings and to proceed with the modeling. Alternatively, you can run the script BR PLS Fit.

The NIPALS Fit Report

Figure 7.32 shows that cross validation has selected the one-factor model, which explains 53% of the variation in the Xs and 31% of the variation in the Ys.

Figure 7.32: PLS Report for One-Factor Fit

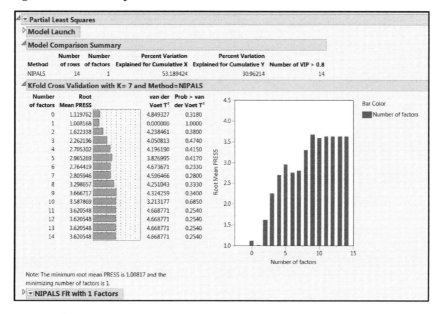

The **Distance Plots** and **T Square Plot** show no obvious outliers or patterns. However, the **Actual by Predicted Plot** report in the **Diagnostics Plots** shows that the predicted values are biased (Figure 7.33). For large observed values, the predicted values tend to be smaller, whereas for small observed values, the predicted values tend to be larger.

Figure 7.33: Actual by Predicted Plots for One-Factor Fit

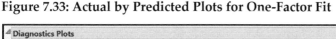
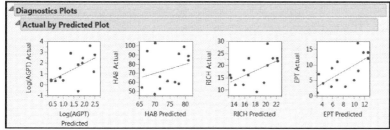

From the red triangle menu at **NIPALS Fit with 1 Factors**, select **VIP vs Coefficients Plots**. The **VIP vs Coefficients for Centered and Scaled Data** plot shows that predictors exhibit the characteristic "V" shape (Figure 7.34). There are several terms below the 0.8 threshold, and 14 above that threshold, as indicated by the **Number of VIP > 0.8** given in the **Model Comparison Summary** (Figure 7.32).

Figure 7.34: VIP vs Coefficients Plot for Centered and Scaled Data

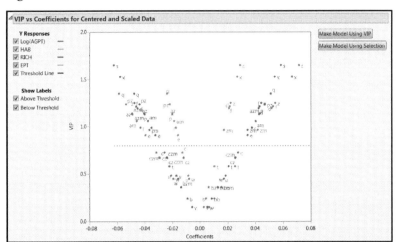

The **VIP vs Coefficients Plot** for the Blue Ridge data shows some differences from the plot for the entire Savannah River basin. The behavior of s and x, which deal with slope, is similar. But notice that q, the percent wetlands, has the next highest VIP. It has a positive coefficient for Log(AGPT), and negative coefficients for HAB, RICH, and EPT. Other predictors, such as e, which measures erodible soils, have moved up in importance.

A Pruned PLS Model for the Blue Ridge Ecoregion

Model Fit

For consistency with the analysis of the whole Savannah River Basin, we prune the model by selecting terms with VIP > 1. As described earlier, select a region of the **VIP vs Coefficients Plot** by dragging a rectangle to capture the required terms. Or use the **Variable Importance Table** approach described earlier. You should have 11 predictors in all. Then click **Make Model Using Selection**. This selection generates a new Fit Model window, containing 11 predictors, shown in Figure 7.35. You can also obtain this window by running the saved script BR Pruned Fit Model Launch.

Figure 7.35: Fit Model Window for the Pruned Model

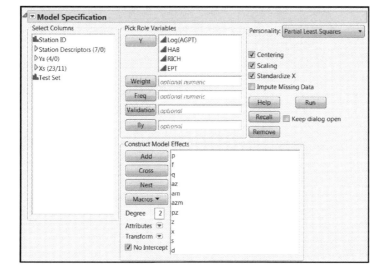

As before, click **Run**, set the random seed to **666**, and then click **Go** to see the results of the cross validation process and the details of the chosen model. Alternatively, run the saved script BR Pruned Fit.

Cross validation selects the one-factor model, which explains 79% of the variation in the **X**s and 31% of the variation in the **Y**s. For comparison, the analysis conducted by Nash and Chaloud (2011) on the same data gives corresponding figures of 94% and 59%, but, as mentioned earlier, their model is restricted to only two of the four **Y**s (HAB and EPT).

The residual plots do not reveal outliers or patterns. However, the **Actual by Predicted Plot** report indicates, as before, that the model is biased. To validate the model against the test data, we need to save the prediction formulas. To this end, select **Save Columns > Save Prediction Formula** from the red triangle menu for the fit. For convenience, place the resulting columns in a column group called Predictions from Pruned Model. To save the prediction formulas and to create the column group directly, run the saved script Add Predictions for BR PLS Pruned Fit.

Comparing Actual Values to Predicted Values for the Test Set

To use only the data that has not been used to build the PLS models for the Blue Ridge ecoregion, run the saved script Use Only Test Data. The script hides and excludes 14 rows.

You can select **Graphs > Graph Builder** to build the requisite plots, or run the saved script Actual by Predicted. The plots are shown in Figure 7.36.

Figure 7.36: Predictions on Test Set for PLS Pruned Model

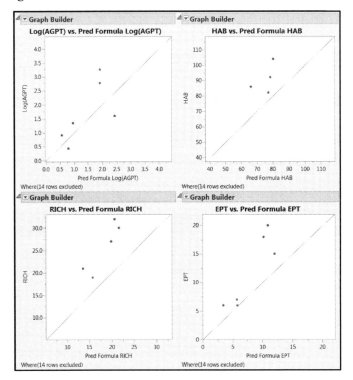

The model appears to be biased for all responses except Log(AGPT). For the other three responses, predicted values are always less than the actual values. The practical impact of this, though, has to be seen in the wider context: For the test data, HAB underpredicts by about 15 units against a range of 22 units, RICH by about 7 units against a range of 13 units, and EPT by about 4 units against a range of 14 units. Note, however, that these conclusions are based on a very small number of points.

This bias is not unexpected. In this example, the variation in the **X**s is well explained, whereas the variation in the **Y**s is not as well explained. The model seems to model the **X**s more accurately than the **Y**s. But such is the nature of PLS—it models both the **X**s and the **Y**s. As mentioned earlier, we deal further with the question of PLS bias in Appendix 2.

Conclusion

We have seen how to fit PLS models to the water quality data featured in Nash and Chaloud (2011). The objective was to relate landscape variables obtained from remote sensing data to measures of water quality obtained from laboratory assessment of samples gathered from field locations. We fit a model for the whole Savannah River Basin and a model for only the Blue Ridge ecoregion. In both cases we used cross validation to select an initial model, and VIPs and coefficients to prune the initial model. We then applied the pruned models to a hold-out set of test data, and found that in many cases the results are biased, but still potentially useful.

The bias in the fits can be mitigated to some extent by fitting each response individually. We encourage you to explore this approach to the analysis using the script ComparePLS1andPLS2 presented in Appendix 2.

Ecology is a specialized topic, and our analysis, out of necessity and expediency, does not emphasize the meaning of the variables used. Although our approach differs in detail from that in Nash and Chaloud (2011), our results conform to their commentary on the major aspects. For example, consider Figure 7.37, which shows the **Coefficients Plot** for the one-factor pruned PLS Blue Ridge ecosystem model for all 20 Blue Ridge observations. (The plot is given by the script Coefficient Plots.) You can relate this plot directly to the discussion in Section 5 of Nash and Chaloud (2011). Note how Log(AGPT) is differentiated from HAB, RICH, and EBT.

Figure 7.37: Coefficient Plot for Blue Ridge PLS Pruned Model

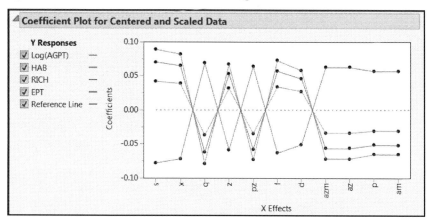

Depending on the kind of data with which you work, you might feel that the predictive power of PLS is a little disappointing in this setting. But you need to view the utility of PLS in a context that is broader than prediction. As you have seen, PLS can provide insight on relationships between the predictors and the responses. In the context of prediction, you must evaluate PLS relative to other analysis approaches commonly used in your area, and also to the information content of the data being analyzed.

Nash and Chaloud (2011) are positive about the practical value of their specific results:

"In both the preliminary and refined models for the whole basin, associations among water biota and landscape variables, largely conform to known ecological processes. In each case the dominant landscape variable corresponds to a critical aspect of the ecoregion: forest in the evergreen forest-dominated Blue Ridge, wetland in the transitional Piedmont, and row crops in the agriculture-dominated Coastal Plain."

They are also positive about the relevance of PLS to similar types of problems:

"The results indicate PLS may prove to be a valuable statistical analysis tool for ecological studies. The data sets used in our analysis contain limitations typical of ecological studies: a small number of sampling sites, a large number of variables, missing values, low signal-to-noise ratio, differences in spatial extent, and different collection methodologies between the field-collection surface water samples and the remote sensing-derived landscape variables. The PLS methodology is less sensitive to these limitations than other statistical methods . . . Univariate-multiple regression analyses with these data sets will not reveal a distinctive pattern of association due to a weak correlation. Summarizing information in the predictor variables by reduction into a few variables . . . makes PLS more suitable in a multivariate context than other, more commonly used, multivariate methods."

Baking Bread That People Like

Background 183
The Data 184
    Data Table Description 184
    Missing Data Check 186
The First Stage Model 187
    Visual Exploration of Overall Liking and Consumer Xs 187
    The Plan for the First Stage Model 189
    Stage One PLS Model 190
    Stage One Pruned PLS Model 195
    Stage One MLR Model 197
    Comparing the Stage One Models 200
The Second Stage Model 202
    Visual Exploration of Ys and Xs 202
    Stage Two PLS Model 207
    The Partial Least Squares Report 208
    Stage Two MLR Model 212
The Combined Model for Overall Liking 215
    Constructing the Prediction Formula 215
    Viewing the Profiler 218
Conclusion 219

Background

Faced with a competitive landscape, consumer goods companies respond by bringing new, innovative products to market. These might be completely new products or variants of existing products that are expected to be more successful than their predecessors.

Extensive consumer research and testing, usually considered part of marketing research, is used to understand the preferences, attitudes, and behaviors of customers and potential customers.

For the technical evaluation of food and drink products, the science of sensory analysis is of critical importance. Sensory analysis applies experimental design and statistical methodology to data relating to the human senses of sight, smell, taste, touch, and hearing. Most consumer goods companies have departments dedicated to sensory analysis, and those that don't usually outsource this important capability.

Both consumer research and sensory analysis constitute considerable bodies of knowledge, and the application of statistical principles and methods in these areas is well established. For introductions to the heavily used statistical approaches, see *Statistics in Market Research* (Chakrapani 2004) and *Sensory Evaluation of Food: Statistical Methods and Procedures* (O'Mahony 1986).

In this chapter, we explore the use of PLS in modeling the relationships between consumer test results on various types of bread, and sensory panel results on the same products. A useful model would allow us to use sensory results to predict likely acceptance by consumers, enabling us to design new desirable products by endowing them with preferable sensory characteristics. In addition, we might be able to determine which characteristics are particularly important to consumer desirability, enabling us to modify these characteristics to produce formulations that are more successful than those currently marketed.

The Data

Data Table Description

Our data are from a large commercial bakery that wanted to understand how sensory results influence consumer desirability. The company's interest focused on 24 types of bread. It conducted both consumer and sensory studies of their bread types. (This example uses data kindly supplied to us by David Rose of SAS.)

For the consumer study, 50 participants were chosen at random from a specific demographic grouping to form a sample representative of the target market. For each type of bread, each participant rated seven attributes on a monotonic scale from 0 to 9 where higher ratings indicate higher desirability. For each type of bread, the scores for each attribute were averaged over all the consumers who took part in the study. These average consumer ratings form the basis of our first analysis.

The company also employed an expert taste panel to assess 60 sensory characteristics for these same 24 breads. Each expert rated every characteristic on a 0 to 9 scale three times. The resulting scores for each characteristic were averaged over all experts and replicates, giving an average sensory score for each type of bread. These 60 sensory ratings are used in our second analysis.

Click on the link in the master journal to open the data table **Bread.jmp**, shown in Figure 8.1. This table contains 24 rows, one for each type of bread studied, and 69 columns. In preparing the table for analysis, all but three of the 69 columns have been arranged in column groups. Specifically, the table contains the following:

- A column called **Bread** that uniquely identifies each type of bread. Note that the bakery sometimes makes more than one type of bread under the same brand.

- A column called **Overall Liking**, the overall affinity that consumers expressed for a given type of bread.

- A group of six columns called **Consumer Xs** that contains the results of the consumer tests for specific aspects of each bread type. In our first analysis, we focus on modeling **Overall Liking** in terms of these columns.

- Five column groupings called **Appearance, Aroma, Aftertaste, Flavor** and **Mouthfeel**. These result from the taste panel's analysis of the 60 sensory characteristics, grouping variables by aspects of the sensory data. Note that the first two characters of each column name indicate the grouping. For example, **Ar Malt** is contained within the **Aroma** grouping and **At Bitter** within the **Aftertaste** grouping. These columns are used as **X**s in a second analysis to model the appropriate columns in **Consumer Xs**.

- A column called **Row State** that holds markers and colors used in the analysis. The colors are determined by the value of **Overall Liking**, with green indicating high values and red indicating low values.

Figure 8.1: Partial View of Bread.jmp

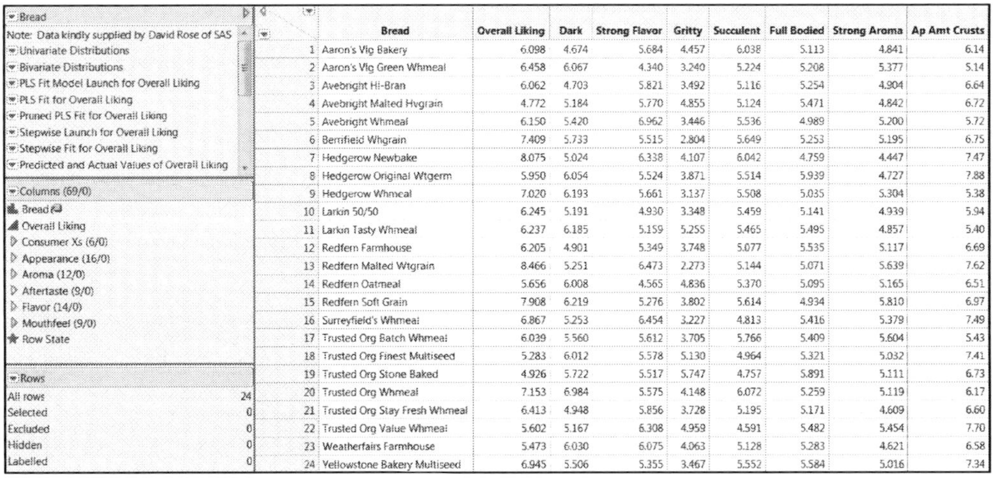

Bread
Note: Data kindly supplied by David Rose of SAS
Univariate Distributions
Bivariate Distributions
PLS Fit Model Launch for Overall Liking
PLS Fit for Overall Liking
Pruned PLS Fit for Overall Liking
Stepwise Launch for Overall Liking
Stepwise Fit for Overall Liking
Predicted and Actual Values of Overall Liking

Columns (69/0)
Bread
Overall Liking
Consumer Xs (6/0)
Appearance (16/0)
Aroma (12/0)
Aftertaste (9/0)
Flavor (14/0)
Mouthfeel (9/0)
Row State

Rows	
All rows	24
Selected	0
Excluded	0
Hidden	0
Labelled	0

	Bread	Overall Liking	Dark	Strong Flavor	Gritty	Succulent	Full Bodied	Strong Aroma	Ap Amt Crusts
1	Aaron's Vlg Bakery	6.098	4.674	5.684	4.457	6.038	5.113	4.841	6.14
2	Aaron's Vlg Green Whmeal	6.458	6.067	4.340	3.240	5.224	5.208	5.377	5.14
3	Avebright Hi-Bran	6.062	4.703	5.821	3.492	5.116	5.254	4.904	6.64
4	Avebright Malted Hvgrain	4.772	5.184	5.770	4.855	5.124	5.471	4.842	6.72
5	Avebright Whmeal	6.150	5.420	6.962	3.446	5.536	4.989	5.200	5.72
6	Berrifield Whgrain	7.409	5.733	5.515	2.804	5.649	5.253	5.195	6.75
7	Hedgerow Newbake	8.075	5.024	6.338	4.107	6.042	4.759	4.447	7.47
8	Hedgerow Original Wtgerm	5.950	6.054	5.524	3.871	5.514	5.939	4.727	7.88
9	Hedgerow Whmeal	7.020	6.193	5.661	3.137	5.508	5.035	5.304	5.38
10	Larkin 50/50	6.245	5.191	4.930	3.348	5.459	5.141	4.939	5.94
11	Larkin Tasty Whmeal	6.237	6.185	5.159	5.255	5.465	5.495	4.857	5.40
12	Redfern Farmhouse	6.205	4.901	5.349	3.748	5.077	5.535	5.117	6.69
13	Redfern Malted Wtgrain	8.466	5.251	6.473	2.273	5.144	5.071	5.639	7.62
14	Redfern Oatmeal	5.656	6.008	4.565	4.836	5.370	5.095	5.165	6.51
15	Redfern Soft Grain	7.908	6.219	5.276	3.802	5.614	4.934	5.810	6.97
16	Surreyfield's Whmeal	6.867	5.253	6.454	3.227	4.813	5.416	5.379	7.49
17	Trusted Org Batch Whmeal	6.039	5.560	5.612	3.705	5.766	5.409	5.604	5.43
18	Trusted Org Finest Multiseed	5.283	6.012	5.578	5.130	4.964	5.321	5.032	7.41
19	Trusted Org Stone Baked	4.926	5.722	5.517	5.747	4.757	5.891	5.111	6.73
20	Trusted Org Whmeal	7.153	6.984	5.575	4.148	6.072	5.259	5.119	6.17
21	Trusted Org Stay Fresh Whmeal	6.413	4.948	5.856	3.728	5.195	5.171	4.609	6.60
22	Trusted Org Value Whmeal	5.602	5.167	6.308	4.959	4.591	5.482	5.454	7.70
23	Weatherfairs Farmhouse	5.473	6.030	6.075	4.063	5.128	5.283	4.621	6.58
24	Yellowstone Bakery Multiseed	6.945	5.506	5.355	3.467	5.552	5.584	5.016	7.34

In the **Columns** panel, you see an icon next to Bread that resembles a tag. This icon indicates that Bread has been designated as a labeling variable. When you click on a point in a plot, the type of Bread for that point appears. (To assign the label role, right - click on a column name in the **Columns** panel and select **Label/Unlabel**.)

As implied earlier, the interpretation of the measured variables suggests a two-step, hierarchical, approach to the modeling: First, we model Overall Liking in terms of the other six consumer ratings, identifying a subset of these **X**s with explanatory power. Then, treating these key variables as **Y**s, we build a second model involving the associated sensory panel results as **X**s. We then implicitly model Overall Liking in terms of the relevant underlying sensory panel results by combining the two explicit models.

Missing Data Check

First we investigate the pattern of missing data. Select **Tables > Missing Data Pattern**. Then select all columns and column groups other than Row State, click **Add Columns**, and then click **OK**. The report indicates that none of the 24 rows have missing values in any of the columns. So, we can exploit the information in all the rows without being concerned about how to handle missing values.

The First Stage Model

Visual Exploration of Overall Liking and Consumer Xs

For the first stage model, **Y** is Overall Liking and the **X**s are the columns in the column group Consumer Xs. Let's look at the univariate distribution of each variable. (The saved script is Univariate Distributions.)

1. Select **Analyze > Distribution**.
2. Select Overall Liking and the Consumer Xs column group from the **Select Columns** list, and click **Y, Columns**.
3. Select the **Histograms Only** check box in the lower left of the window.
4. Click **OK**.

Select the top three bars of the histogram for Overall Liking by clicking on them while holding down the Shift key. You see the report shown in Figure 8.2.

Figure 8.2: Univariate Distributions of Consumer Study Results

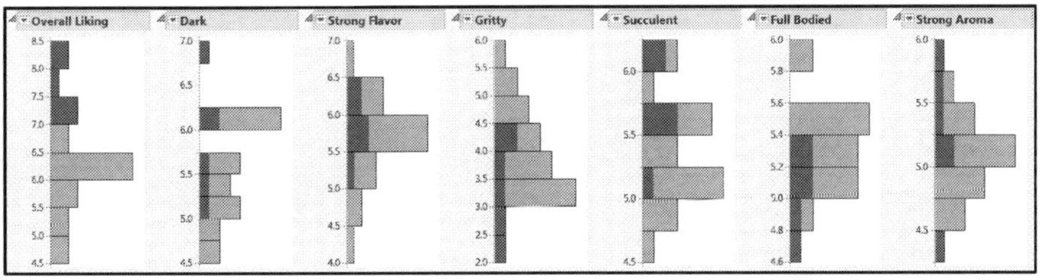

The distributions seem relatively well behaved. The highlighting shows the distribution of the consumer preferences for the types of bread that are liked best. Gritty and Full Bodied both have generally low values for these breads, but the patterns for other characteristics are not as clear.

To explore the relationships between variables two at a time, complete the following steps, or, run the script Bivariate Distributions:

1. Select **Analyze > Multivariate Methods > Multivariate**.
2. Enter Overall Liking and the columns in Consumer Xs as **Y, Columns**.
3. Click **OK**.

The report gives **Correlations** and a **Scatterplot Matrix**. In the **Scatterplot Matrix**, the points that are not currently selected in the table Bread.jmp are muted. To undo the selection, click in some blank space within the matrix. The **Scatterplot Matrix** is shown in Figure 8.3.

Figure 8.3: Scatterplot Matrix for Consumer Study Results

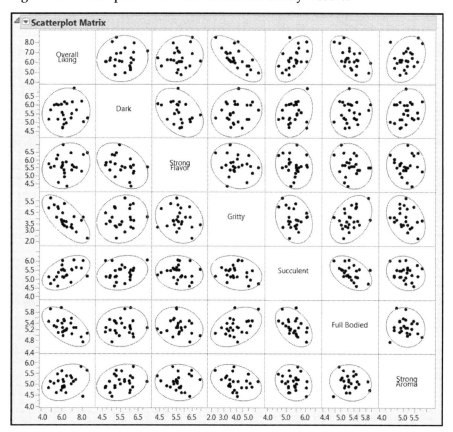

The confidence ellipses that are shown by default in the scatterplot matrix are based on the assumption that the data follow a multivariate normal distribution. This assumption might not apply for our consumer ratings, so these ellipses serve only as a useful guide to the eye. Position your mouse pointer near the points outside the ellipses. Tooltips appear that show the type of bread and the values of the two variables for that bread type. No one type of bread appears consistently as an outlier in these bivariate distributions, and no single point appears to be too distant from the cloud of points.

The **Correlations** report is shown in Figure 8.4. The correlations are only moderate; the correlation with the largest absolute value is –0.6764. Nonetheless, Overall Liking is correlated with some of the Xs, and some of the Xs are mutually correlated. (For example, Full Bodied and Succulent have a correlation of –0.4311.)

Figure 8.4: Correlations between Consumer Study Variables

Correlations	Overall Liking	Dark	Strong Flavor	Gritty	Succulent	Full Bodied	Strong Aroma
Overall Liking	1.0000	0.0649	0.1640	-0.6764	0.4638	-0.5501	0.2921
Dark	0.0649	1.0000	-0.3672	0.1437	0.2082	0.0882	0.2228
Strong Flavor	0.1640	-0.3672	1.0000	-0.1685	-0.1172	-0.1548	-0.0463
Gritty	-0.6764	0.1437	-0.1685	1.0000	-0.2092	0.3758	-0.2974
Succulent	0.4638	0.2082	-0.1172	-0.2092	1.0000	-0.4311	-0.1674
Full Bodied	-0.5501	0.0882	-0.1548	0.3758	-0.4311	1.0000	-0.0895
Strong Aroma	0.2921	0.2228	-0.0463	-0.2974	-0.1674	-0.0895	1.0000

The Plan for the First Stage Model

We make the assumption that the specific characteristics represented in Consumer Xs are connected with Overall Liking. We are interested in determining which of these characteristics are most influential in this relationship. In other words, we are interested in *variable selection*.

There are many statistical approaches to variable selection. Generally speaking, PLS should be used cautiously for this purpose. (See Appendix 2.) Given that PLS is often used when variables are heavily interrelated, variable selection can result in choosing variables that don't make sense from an explanatory perspective. But the fact remains that PLS is often used for variable selection. In the final analysis, a useful model is the goal and the model's utility is dependent on both the data and the wider objectives and context of the analysis. Our advice is to seek confirmation using alternative approaches when possible.

Recall that there are 24 rows in Bread.jmp, Overall Liking is a single Y, and there are only six variables in Consumer Xs. So in addition to using PLS, we can also use multiple linear regression (MLR) to identify the Xs that drive Overall Liking. In the case of MLR, we will use stepwise regression for model selection.

We will attempt to identify the important consumer predictors using both methods: PLS and stepwise regression. To identify important predictors using PLS, we first fit a PLS model using all predictors, and then use the VIPs and model coefficients to select predictors. Next we perform variable selection based on MLR using a stepwise approach. We compare the results of the two methods to determine a final set of predictors.

In terms of validation, we do the following:

- PLS: To avoid fitting noise, given that we have relatively few rows available, we use leave-one-out cross validation. This choice makes good use of the data and, with small data sets, runs quickly. (In case this option isn't available to you in your version of JMP, in the PLS Model Launch control panel, you can specify leave-one-out cross validation by setting the **Number of Folds** equal to the number of rows, here, 24).
- Stepwise: There are several alternative stopping rules for the stepwise personality. We will use **K-Fold Crossvalidation** with k = 24 for consistency with our PLS choice of leave-one-out cross validation.

Although you generally need to specify a **Random Seed** to reproduce cross validation results, for leave-one-out cross validation specifying a seed is unnecessary, because there is only one way to define the folds.

Stage One PLS Model

We describe the JMP Pro PLS launch through Fit Model. (The script is PLS Fit Model Launch for Overall Liking.) If you are running JMP, you can launch the platform by selecting **Analyze > Multivariate Methods > Partial Least Squares** and adapt the following steps.

1. Select **Analyze > Fit Model**.
2. Enter Overall Liking as **Y**.
3. Enter all the columns in the group Consumer Xs in the **Construct Model Effects** list.
4. Select **Partial Least Squares** as the **Personality**.

   Your window should appear as shown in Figure 8.5.

5. Click **Run**.

Figure 8.5: Fit Model Window

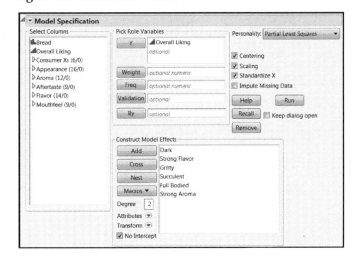

In the PLS Model Launch control panel, accept the default **Model Specification** of **NIPALS**, but either change the **Validation Method** to **Leave-One-Out**, or change the **Number of Folds** for **KFold** validation to **24**. Click **Go**. (The saved script is PLS Fit for Overall Liking.)

The Partial Least Squares Report

Clicking **Go** adds three report sections (Figure 8.6): A **Model Comparison Summary** report, which is updated when new fits are performed, a **Leave-One-Out (or KFold) Cross Validation** report, and a **NIPALS Fit with 1 Factors** report that details the fit chosen through the cross validation procedure.

Figure 8.6: PLS Report for One-Factor Model

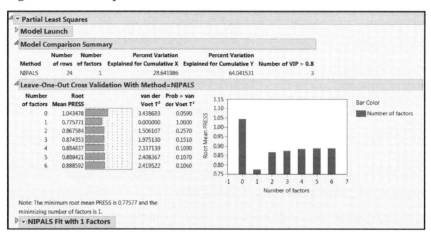

The **Leave-One-Out (or KFold) Cross Validation** report shows that **Root Mean PRESS** is minimized with one factor. The **Prob > van der Voet** values indicate that PRESS for a single-factor model differs significantly from PRESS for a no-factor model. Keep in mind that, because the van der Voet p-values are obtained by simulation, your **Prob > van der Voet** values might differ from those shown in Figure 8.6.

The **Model Comparison Summary** indicates that the single factor accounts for about 29% of the variation in the **X**s and 64% of the variation in the **Y**s. The last column, **Number of VIP > 0.8**, indicates that only 3 of the 6 customer preference predictors are influential in determining the single factor. This suggests some potential for refining the model by dropping some of the predictors.

The NIPALS Fit Report

Now let's look at the report for the fit, **NIPALS Fit with 1 Factors** (Figure 8.7). The **X-Y Scores Plots** report shows reasonable correlation between the X and Y scores. In a good PLS model, the first few factors should show a high correlation between the X and Y scores and, although there is some scatter here, the correlation is reasonable.

The **Percent Variation Explained** report displays the variation in the **X**s and **Y**s that is explained by each factor. The **Cumulative X** and **Cumulative Y** values must agree with the corresponding figures in the **Model Comparison Summary**. Because of the small to moderate correlations among the **X**s, we are not surprised that the model explains only 29% of the variation in the **X**s.

The two **Model Coefficients** reports give the estimated model coefficients for predicting the Ys. Recall that the **Model Coefficients for Centered and Scaled Data** report gives coefficients that apply when the Xs and Ys have been transformed to have mean zero and standard deviation one. If the **Standardize X** option is selected, the **Model Coefficients for Original Data** report gives coefficients for the model expressed in terms of the standardized predictors. If the **Standardize X** option is not selected, the report gives the coefficients for the model in terms of the raw data values. In both cases, this second set of model coefficient estimates is of secondary interest.

Figure 8.7: NIPALS Fit Report

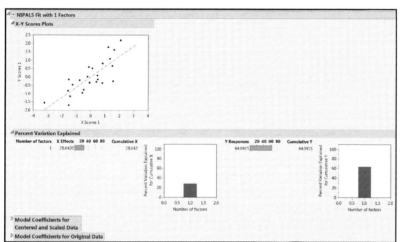

Select **Diagnostics Plots** from the **NIPALS Fit with 1 Factors** red triangle menu to obtain the plots shown in Figure 8.8. These plots indicate no issues of note.

Figure 8.8: Diagnostics Plots

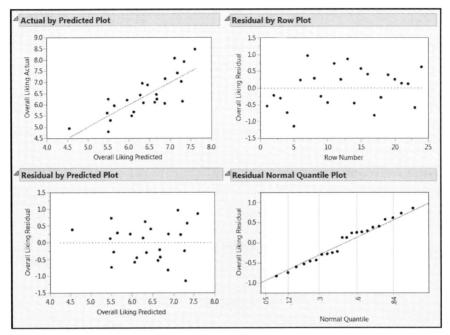

To compare this first model with others we build, select **Save Columns > Save Prediction Formula** from the red triangle menu for the NIPALS fit. This adds a new formula column, called Pred Formula Overall Liking, to the Bread.jmp data table.

To explore possibilities for pruning the model by removing Xs, select the **VIP vs Coefficients Plots** option from the **NIPALS Fit with 1 Factors** red triangle menu (Figure 8.9). Note there is an overall "V" shape, indicating that the terms that contribute to the regression also are important in PLS dimensionality reduction.

Figure 8.9: VIP versus Coefficients Plot

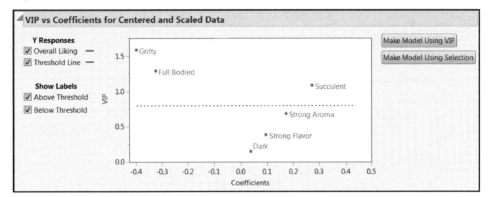

The two buttons to the right of the plot in Figure 8.9 are convenient for producing a pruned model. Click the **Make Model Using VIP** button to use the default VIP threshold of 0.8. This selection produces a Fit Model launch window containing the terms with VIP values of 0.8 or higher: Gritty, Succulent, and Full Bodied.

Stage One Pruned PLS Model

Model Fit

To fit the pruned model using the chosen terms, click **Run** in the Fit Model launch window. In the resulting PLS Model Launch control panel, accept the default **Model Specification** of **NIPALS**, but as for the full model, select **Leave-One-Out** as the **Validation Method**. (Or, use **KFold** with the **Number of Folds** equal to 24.) Click **Go**. (The script is Pruned PLS Fit for Overall Liking.)

Cross validation selects a one-factor model, producing the report shown in Figure 8.10.

Figure 8.10: PLS Report for One-Factor Pruned Model

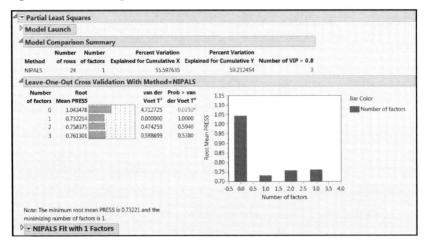

The one-factor model based on our variable selection explains 56% of the variation in the **X**s and 59% of the variation in **Y**, Overall Liking, compared to the corresponding figures of 29% and 64%, respectively, for the full model.

Open the **NIPALS Fit with 1 Factors** report to review the **X-Y Scores Plots**, shown in Figure 8.11.

Figure 8.11: NIPALS Fit Report for the Pruned Model

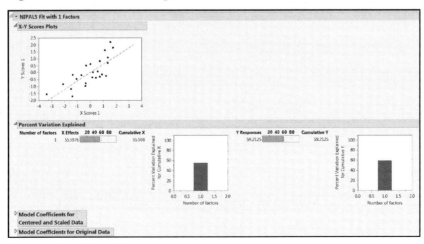

Diagnostics

From the red triangle menu next to the model fit, select **Diagnostics Plots** to produce Figure 8.12.

The **Actual by Predicted Plot** report shows a somewhat unbiased fit with some variation. There is a suggestion that larger values of Overall Liking are being slightly underpredicted.

The **Residual by Predicted Plot** shows no anomalies or patterns. The **Residual Normal Quantile Plot** shows some non-randomness in that points are systematically below and above the red line as we move from left to right. This effect is a little more pronounced than for the full model in Figure 8.8. However, this issue is not serious and we do not pursue it.

Figure 8.12: Diagnostics Plots for the Pruned Model

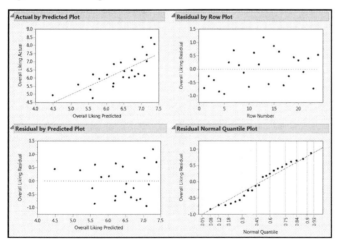

To compare this pruned model with other models, select **Save Columns > Save Prediction Formula** from the red triangle menu for the fit to generate a new formula column in Bread.jmp. The new column is called Pred Formula Overall Liking 2.

Stage One MLR Model

We use the stepwise personality to perform variable reduction based on multiple linear regression fits. Complete the following steps. (Or, run the script Stepwise Launch for Overall Liking.)

1. Select **Analyze > Fit Model**.
2. Enter Overall Liking as **Y**.
3. Enter all the columns in the group Consumer Xs into the **Construct Model Effects** list.
4. Select **Stepwise** as the **Personality**.

   Your window should appear as shown in Figure 8.13.
5. Click **Run.**

Figure 8.13: Fit Model Window for Stepwise Multiple Linear Regression

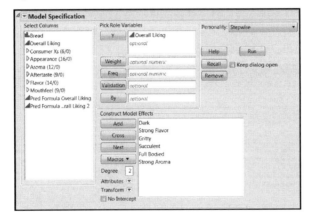

6. Select **K-Fold Crossvalidation** from the report's red triangle menu.
7. Enter a value of **24** for **k** in the menu that opens.
8. Click **OK.**

   This results in the **Stepwise Regression Control** report shown in Figure 8.14. (The script Stepwise Fit for Overall Liking produces this report.)
9. Click **Go**.

Figure 8.14: Control Panel for Stepwise Multiple Linear Regression

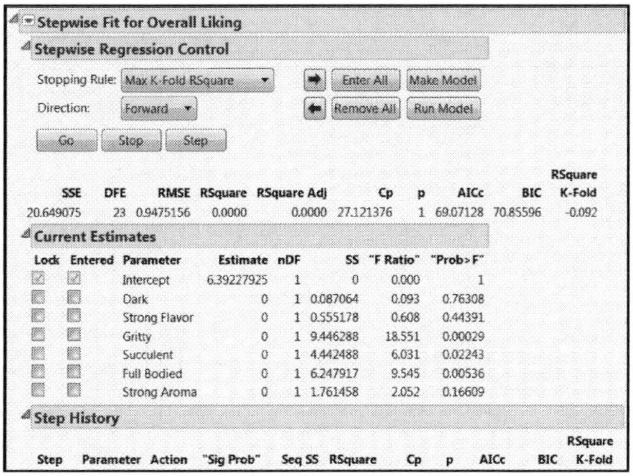

The variable reduction process runs until the stopping rule is triggered. The variables Gritty, Succulent, and Full Bodied are selected. These are precisely the same terms that appear in the pruned PLS model.

To fit a linear regression model to only these three terms, click the **Run Model** button. In the resulting report, from the red triangle menu next to **Response Overall Liking**, select **Save Columns > Prediction Formula** (Figure 8.15). This saves the prediction formula so that we can compare it to the two formulas obtained using PLS. The new column is called Pred Formula Overall Liking 3.

Figure 8.15: Multiple Linear Regression Fit with Terms Selected by Stepwise

Fit Group

Response Overall Liking

Summary of Fit

RSquare	0.609142
RSquare Adj	0.550513
Root Mean Square Error	0.63525
Mean of Response	6.392279
Observations (or Sum Wgts)	24

Analysis of Variance

Source	DF	Sum of Squares	Mean Square	F Ratio
Model	3	12.578221	4.19274	10.3898
Error	20	8.070854	0.40354	Prob > F
C. Total	23	20.649075		0.0002*

Parameter Estimates

Term	Estimate	Std Error	t Ratio	Prob>|t|
Intercept	9.9530172	4.054378	2.45	0.0234*
Gritty	-0.599628	0.169897	-3.53	0.0021*
Succulent	0.5903326	0.370085	1.60	0.1264
Full Bodied	-0.822699	0.554456	-1.48	0.1535

Effect Tests

Effect Details

Comparing the Stage One Models

A Graphical Comparison

To facilitate interpreting subsequent reports, rename the three saved prediction formula columns as follows:

- Pred Formula Overall Liking as Pred Formula Overall Liking – PLS.
- Pred Formula Overall Liking 2 as Pred Formula Overall Liking – PLS Pruned.
- Pred Formula Overall Liking 3 as Pred Formula Overall Liking – Regression.

To rename a column, *slowly* double-click on the column name in the **Columns** panel and enter the new name.

Now select **Graph > Graph Builder**. Drag Overall Liking to the **X** area. Drag each of the three prediction formula columns individually to the **Y** area, depositing each at the top of the axis to produce individual displays. Your plot should be similar to that shown in Figure 8.16. (This graph is one of two produced by the script Predicted and Actual Values of Overall Liking, which also generates the three prediction formulas.)

Figure 8.16: Comparison of Fitted versus Actual Values of Overall Liking

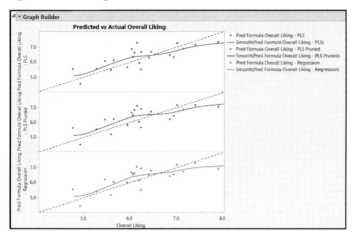

The diagonal dotted line is the line of perfect agreement. You can add this line by right-clicking on the respective white region in each graph box, selecting **Customize**, and then clicking the **+** sign. In the text box, enter the following expression:

> **Line Style("Dashed"); Y Function(x, x);**

Click **OK**.

These lines serve as guides, enabling you to gauge the quality of each fit and assess the differences between fits. These fits appear quite similar, but we stress that this might not always be the case when comparing PLS and MLR.

Comparison via the Profiler

For another view, select **Graph > Profiler**. Assign the three formula columns to the **Y, Prediction Formula** role and click **OK**. This produces the report shown in Figure 8.17. (This graph is one of two produced by the script Predicted and Actual Values of Overall Liking.)

Figure 8.17: Simultaneous Profiling of Three Stage One Models of Overall Liking

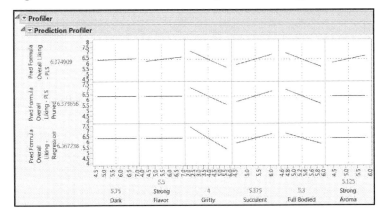

In the second and third profiler rows, the cells for the three variables, Dark, Strong Flavor, and Strong Aroma, show perfectly horizontal lines. These variables were not included in the pruned PLS or reduced regression models. They were included in our first PLS model, which is the model represented in the top row of the plot.

Manipulating the dotted vertical lines produces different profiles of Overall Liking within the six-dimensional or three-dimensional **X** space. By varying the settings, you see that the model predictions are in good agreement. In fact, predictions for the bottom two models are virtually identical for practical purposes.

Based on the stage one analysis, the customer preference columns Gritty, Succulent, and Full Bodied become the **Y**s for our stage two analysis. In that stage, we model these three variables as **Y**s in terms of the expert taste panel results.

The Second Stage Model

Visual Exploration of Ys and Xs

In this analysis we have three **Y**s (Gritty, Succulent, and Full Bodied), 60 **X**s (in 5 groups), and 24 rows. Because we have more predictors than rows, it is not possible to use MLR.

To help make subsequent displays more meaningful, we color the rows in Bread.jmp according to the values of Overall Liking. To color the rows, complete the following steps:

1. Select **Rows > Color or Mark by Column**.
2. Select Overall Liking from the **Columns** list.

3. From the **Colors** list, select **Green to Black to Red**.

4. Select the **Reverse Scale** check box.

   Reversing the scale causes small values of Overall Liking to be colored red and large values green, which is more intuitive.

5. Click **OK**.

6. Select all the rows, right-click in the highlighted area, select **Markers**, and then select the open circle marker.

The previous two steps update the row states of Bread.jmp. Alternatively, you can apply these row states using the Row State column, where they are already saved. To apply row states from the Row State column, click on the red star next to the Row State column name in the **Columns** panel and select **Copy to Row States**.

We can conveniently assess the pattern of variation between Gritty, Succulent, and Full Bodied using a 3-D Scatterplot. Select **Graph > Scatterplot 3D**, enter the columns Gritty, Succulent, and Full Bodied as **Y, Columns**, and then click **OK**. From the report's red triangle menu, select **Normal Contour Ellipsoids**. In the menu that appears, set **Coverage** to 0.95 and click **OK**. After a little rescaling of the three axes, your plot should appear as in Figure 8.18. Alternatively, run the script Scatterplot 3D of Taste Panel Ys to obtain this plot.

Figure 8.18: Pattern of Variation for Three Taste Panel Ys

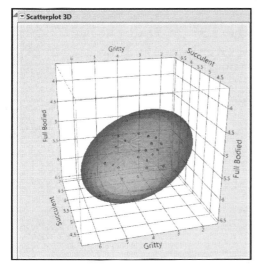

Rotating the cube shows that the cloud of points has a compact distribution, and that the three **Y**s are moderately correlated. (See also Figure 8.4.)

To review the univariate distribution of the 60 **X**s, select **Analyze > Distribution**, add the five column groups Appearance, Aroma, Aftertaste, Flavor, and Mouthfeel to the **Y, Columns** role, select the **Histograms Only** check box, and then click **OK**. Alternatively, generate Figure 8.19 by running the saved script Univariate Distributions of Taste Panel Xs. Note that some of the columns, such as Ap Black Seeds, seem to have a bimodal distribution. To see if this is a common feature across several columns, select the top two bars of the histogram for Ap Black Seeds to produce the highlighting shown. (Click and hold down the Shift key to select both bars.) Ap White Husks clearly exhibits this effect too.

Figure 8.19: Univariate Distributions of Taste Panel Xs (Partial View)

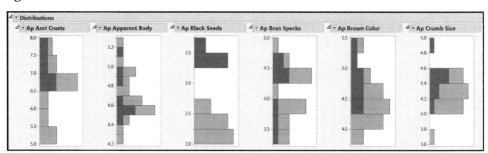

To visualize the distributions of the **X**s two at a time, select **Analyze > Multivariate Methods > Multivariate**, assign the groups Appearance, Aroma, Aftertaste, Flavor, and Mouthfeel to the **Y, Columns** role, and then click **OK**. From the report's red triangle menu, select **Scatterplot Matrix**. Alternatively, run the script Bivariate Distributions of Taste Panel Xs. Be patient—this computation takes a little time. Figure 8.20 shows a partial view of the results.

Figure 8.20: Bivariate Distributions of Taste Panel Xs (Partial View)

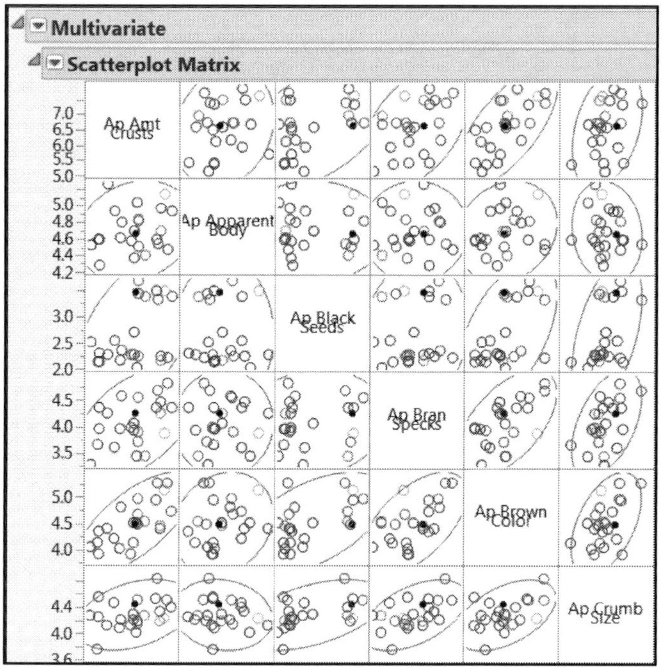

Aside from the bimodality observed already, the points generally form compact groups within the ellipses with no obvious extreme values, and with moderate pairwise correlations. We can investigate the correlation structure further by selecting **Color Maps > Color Map on Correlations** from the **Multivariate** red triangle menu in the report window, producing Figure 8.21. This report is provided if you have run the script Bivariate Distributions of Taste Panel Xs.

Figure 8.21: Pairwise Correlations between Taste Panel Xs

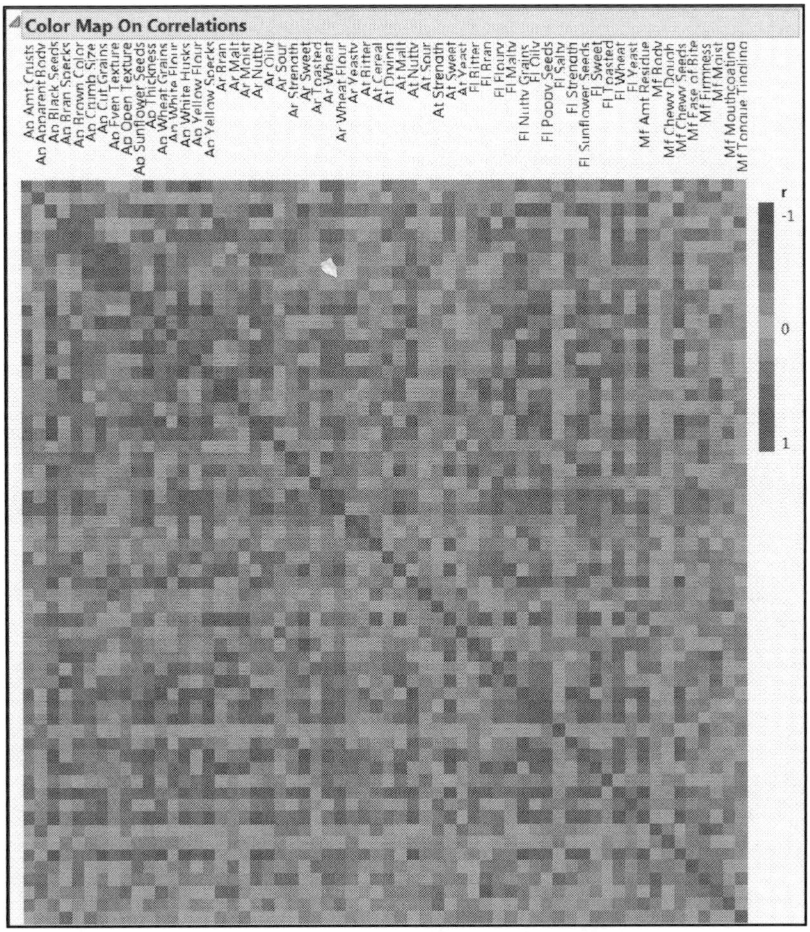

Generally, the correlations are moderate to high, with some evidence that the columns within groups are particularly strongly related.

To investigate the variation between Xs and Ys, select **Graph > Scatterplot Matrix**. Assign Gritty, Succulent, and Full Bodied to the **Y, Columns** role and the groups Appearance, Aroma, Aftertaste, Flavor, and Mouthfeel to the **X** role. From the **Matrix Format** list, select **Square**. Click **OK**. From the report's red triangle menu, select **Density Ellipses > Density Ellipses** to produce Figure 8.22. (The script is Scatterplot Matrix for Taste Panel Xs and Ys.)

Figure 8.22: Bivariate Distributions of Taste Panel Ys and Xs (Partial View)

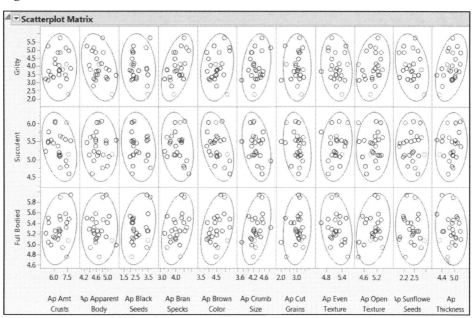

None of the associations look particularly strong, although some are stronger than others. There are no obviously extreme points in any **X** and **Y** combination.

Stage Two PLS Model

Fitting a First Model

We use the JMP Pro PLS launch through Fit Model. (The script is PLS Fit Model Launch for Taste Panel Ys.) Remember that, if you are running JMP, you can launch the platform by selecting **Analyze > Multivariate Methods > Partial Least Squares**.

1. Select **Analyze > Fit Model**.
2. Enter Gritty, Succulent, and Full Bodied as **Y**.
3. Enter all the columns in the groupings Appearance, Aroma, Aftertaste, Flavor, and Mouthfeel as **Model Effects**.
4. Select **Partial Least Squares** as the **Personality**.

   Your window should appear as shown in Figure 8.23.

5. Click **Run**.

Figure 8.23: Fit Model Window

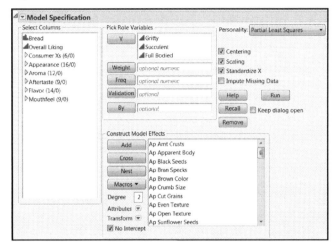

In the PLS Model Launch control panel, accept the default **Model Specification** of **NIPALS**, but set the **Validation Method** to **Leave-One-Out**. Click **Go**. (The script is PLS Fit for Taste Panel Ys.)

The Partial Least Squares Report

The usual results are appended: A **Model Comparison Summary** report, a **KFold Cross Validation** report, and a report entitled **NIPALS Fit with 2 Factors**, detailing the fit chosen through the cross validation procedure (Figure 8.24).

Figure 8.24: PLS Report for Two-Factor Fit

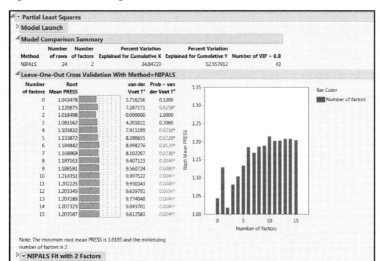

The **Leave-One-Out (or KFold) Cross Validation** report shows that **Root Mean PRESS** is minimized with two factors. The **Prob > van der Voet** values indicate that PRESS for a single-factor model differs significantly from PRESS for a two-factor model. Note that your **Prob > van der Voet** values might differ from those shown in Figure 8.24, as these are obtained by simulation. We conclude that a two-factor model is appropriate.

The **Model Comparison Summary** indicates that two factors account for about 35% of the variation in the **X**s and 53% of the variation in the **Y**s. The last column, **Number of VIP > 0.8**, indicates that 43 of the 60 taste panel preference predictors are influential in determining the factors.

The NIPALS Fit Report

Now let's look at the report for the fit, **NIPALS Fit with 2 Factors** (Figure 8.25). The **X-Y Scores Plots** report shows reasonable correlation between the X and Y scores.

The **Percent Variation Explained** report displays the variation in the **X**s and **Y**s that is explained by each factor. The **Cumulative X** and **Cumulative Y** values must agree with the corresponding figures in the **Model Comparison Summary**. Because the **X**s are not highly correlated, we are not surprised that the model explains only 35% of the variation in the **X**s.

The two **Model Coefficients** reports give the estimated model coefficients for predicting the **Y**s. Recall that the **Model Coefficients for Centered and Scaled Data** report is of primary interest. It gives coefficients that apply when the **X**s and **Y**s have been transformed to have mean zero and standard deviation one.

Figure 8.25: The NIPALS Fit Report

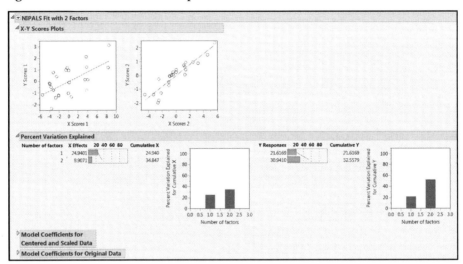

The **Diagnostics Plots** (Figure 8.26) show no obvious issues. (The script PLS Fit for Taste Panel Ys 2 opens to show Figures 8.25, 8.26, and 8.27.)

Figure 8.26: Diagnostics Plots

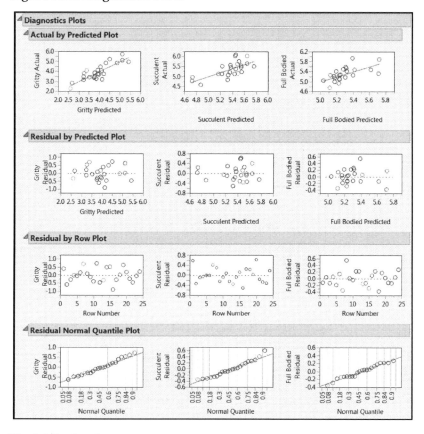

Model Reduction

We would like to narrow the list of **X**s to a manageable number so that we can focus on modifying a select number of characteristics to improve Overall Liking (and sales!). We would also like to use our manageable number of **X**s to predict changes in Overall Liking. To this end, we decide to focus on only those predictors whose coefficients for the centered and scaled data have magnitudes exceeding 0.1 for at least one of the three responses. Admittedly, the value 0.1 is arbitrarily chosen to produce a manageable set of predictors and to allow the use of a multiple regression model.

Select **Coefficients Plots** from the red triangle menu of the **NIPALS Fit** to obtain the plot in Figure 8.27. The horizontal dashed reference lines at −0.1 and +0.1 were added by the script PLS Fit for Taste Panel Ys 2.

Figure 8.27: Coefficient Plot for the Stage Two PLS Model

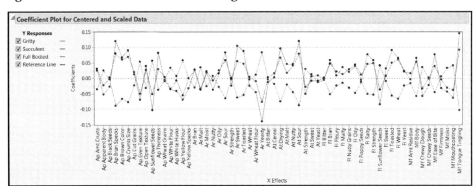

Note that Figure 8.27 shows an interesting pattern in the coefficient values—Full Bodied and Gritty follow a similar pattern, while Succulent shows that same pattern but with the opposite sign. This information alone should be of value in terms of understanding consumer preferences.

Only six **X**s, Ap Bran Specs, Ap Sunflower Seeds, Ar Sweet, Ar Yeasty, At Sour, and Mf Tongue Tingling have coefficients with magnitudes exceeding 0.1. Note that we could not make this, or any other, selection of variables from MLR directly, because multiple regression with the full set of variables is not possible.

At this point, you could construct the regression model that you want to fit as follows: Select **VIP vs Coefficients Plots** from the red triangle **NIPALS Fit with 2 Factors** menu. In the **VIP vs Coefficients for Centered and Scaled Data** plot, select points where **Coefficients** is less than –0.10. Then, holding down the Shift key, select points where **Coefficients** is greater than 0.10. Click **Make Model Using Selection**. Change the **Personality** to **Standard Least Squares**. Or, proceed as described in the next section.

Stage Two MLR Model

Fit a multiple linear regression model to our three **Y**s by completing the following steps. (Alternatively, run the saved script Regression Fit Model Launch.)

1. Select **Analyze > Fit Model**.

2. Enter Gritty, Succulent, and Full Bodied as **Y**.

3. Enter the columns Ap Bran Specs, Ap Sunflower Seeds, Ar Sweet, Ar Yeasty, At Sour, and Mf Tongue Tingling as **Model Effects**.

4. Select **Standard Least Squares** as the **Personality**.

   Your Fit Model window should appear as shown in Figure 8.28.

5. Click **Run.**

Figure 8.28: Fit Model Window for Multiple Linear Regression Model

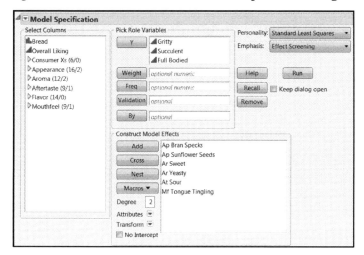

The report, with some closed report sections, is shown in Figure 8.29. There is a report for each of the three responses, followed by a report for the **Prediction Profiler**. The **Actual by Predicted** plots show no anomalies. The report indicates that the overall model for Gritty is significant. The models for Succulent and Full Bodied are not significant, but keep in mind that we are interested in the effects of the predictors on all three responses, and eventually, on Overall Liking.

Figure 8.29: Multiple Linear Regression Report

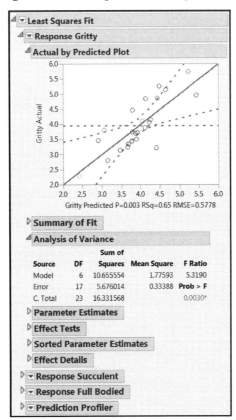

The **Prediction Profiler** is shown at the bottom of the Least Squares Fit report (Figure 8.30). Profiler plots for all three responses are shown. The dotted blue lines show the uncertainty in the fit, and you can drag the vertical dotted red lines to profile the **Y**s as the six-dimensional factor space is cut.

Figure 8.30: Profilers for Taste Panel Ys and Xs

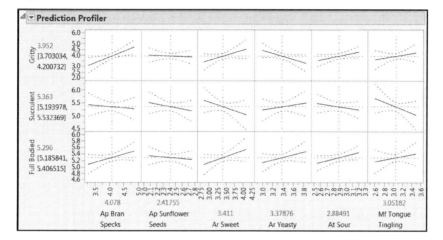

While holding down the Ctrl key, select **Save Columns > Prediction Formula** from the **Response Gritty** red triangle menu. This saves a prediction column in Bread.jmp for each of the three responses. Alternatively, you can add these columns by running the script Add Predictions for Consumer Xs.

The Combined Model for Overall Liking

Constructing the Prediction Formula

The table Bread.jmp now contains six prediction formula columns (Figure 8.31). The first three come from modeling Overall Liking in terms of Gritty, Succulent, and Full Bodied using PLS (twice) and MLR, and the second three come from modeling each of Gritty, Succulent, and Full Bodied in terms of Ap Bran Specs, Ap Sunflower Seeds, Ar Sweet, Ar Yeasty, At Sour, and Mf Tongue Tingling using MLR.

Next we construct a prediction formula for Overall Liking using the two stages of modeling. We use Pred Formula Overall Liking – Regression as our first stage model for Overall Liking. Then, for each of the Gritty, Succulent, and Full Bodied terms that appear in that formula, we substitute the prediction formula for that term obtained from our second stage regression. This gives a formula for Overall Liking expressed in terms of the six key taste panel terms. Note that the model defined by the two-stage formula differs from a single-stage regression model for Overall Liking using the same six terms. The two-stage model connects the taste panel Xs explicitly to the consumer Xs, and then these are connected to Overall Liking.

Figure 8.31: Prediction Formula Columns from First and Second Stage Models

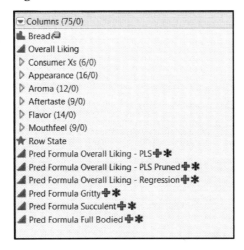

Click the **+** sign to the right of Pred Formula Gritty in the **Columns** panel to reveal the formula editor (Figure 8.32). Right-click in the formula in the editor window. As shown in Figure 8.32, you can **Copy** and **Paste** the formula. Note that the scope of this operation is defined by the red box in the formula editor.

Figure 8.32: Formula for Gritty from Second Stage MLR Model

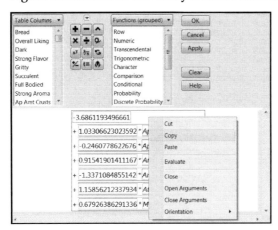

Using this feature, complete the following steps. Alternatively, you can run the script Add Two-Stage Prediction of Overall Liking.

1. Use **Cols > New Column** to make a new column called Prediction of Overall Liking from Taste Panel Xs.

2. Select **Formula** from the **Column Properties** menu.

   This opens a formula editor window.

3. Copy the formula expression from Pred Formula Overall Liking – Regression, as described earlier.

4. Paste that formula into the formula editor for Prediction of Overall Liking from Taste Panel Xs. To do this, you can hold down the Ctrl key and click **v**. Alternatively, double-click inside the red outline. It turns blue. Right-click inside and select **Paste**. Leave this formula editor window open.

5. Copy the formula from the column Pred Formula Gritty.

6. In the Prediction of Overall Liking from Taste Panel Xs formula editor, select **Gritty** in the expression. Make sure that there is a red box surrounding **Gritty**.

7. Paste the formula from Pred Formula Gritty into this red box. You can hold down the Ctrl key and click **v**, or you can right-click inside the red box and select **Paste**.

8. Close the formula editor window for Pred Formula Gritty.

9. Repeat steps 5 through 8 for Succulent and Full Bodied using the appropriate formulas.

   Now the column Two-Stage Prediction of Overall Liking contains a single formula relating Overall Liking to Ap Bran Specs, Ap Sunflower Seeds, Ar Sweet, Ar Yeasty, At Sour, and Mf Tongue Tingling. The complete formula is shown in Figure 8.33.

10. Click **OK** to close the formula editor window for the column Two-Stage Prediction of Overall Liking.

Figure 8.33: Formula for Overall Liking from the Two-Stage Model

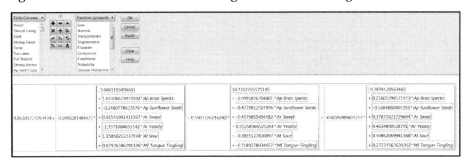

Viewing the Profiler

Now select **Graph > Profiler**. Enter Two-Stage Prediction of Overall Liking as **Y, Prediction Formula**. Click **OK**. (The script is Profiler for Overall Liking.) The profiler is shown in Figure 8.34.

Figure 8.34: Profiler for Overall Liking in Terms of Key Taste Panel Xs

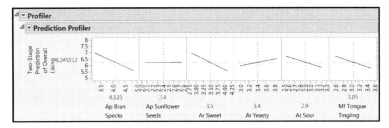

The profiler shows a surprising but not contradictory result, namely that one of our key taste panel Xs (Ap Sunflower Seeds) has little or no effect on Overall Liking. For some insight on why this happens, open the formula editor for Two-Stage Prediction of Overall Liking. Notice the red triangle above the key pad. From the red triangle menu, select **Simplify** (Figure 8.35). The coefficient of Ap Sunflower Seeds in the combined model is close to zero, with a value of 0.004.

Figure 8.35: Simplified Two-Stage Formula for Overall Liking

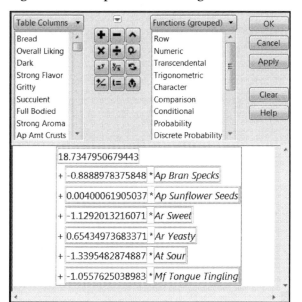

You can investigate how Ap Bran Specs, Ar Sweet, Ar Yeasty, At Sour, and Mf Tongue Tingling affect Overall Liking by interacting with the Profiler and performing informative what-if scenarios using the built-in simulator. (To access the simulator, select **Simulator** from the **Prediction Profiler** red triangle menu.)

Conclusion

We have seen how to use PLS and MLR in combination to effectively model relationships among overall liking, 6 consumer-testing results, and 60 sensory panel results on 24 types of bread. Although its use for variable selection is sometimes controversial, PLS can be viewed as another tool for selecting important variables. It is of particular value when MLR is not viable. But the ultimate test of whether a model is useful is how well it survives contact with new data, and no matter how it was built, models should be continually challenged in this way to avoid pitfalls.

The new understanding conveyed by the composite model enables the bakery to focus on a small number of sensory characteristics in developing new breads that consumers rate more highly. As well as providing useful direction as to how the new bread should be formulated and baked, the bakery might also be able to reduce the cycle time and effort required for sensory testing.

Appendix: Technical Details

Ground Rules	222
The Singular Value Decomposition of a Matrix	222
Definition	222
Relationship to Spectral Decomposition	223
Other Useful Facts	223
Principal Components Regression	223
The Idea behind PLS Algorithms	224
NIPALS	225
The Algorithm	225
Computational Results	228
Properties of the NIPALS Algorithm	231
SIMPLS	237
Optimization Criterion	237
Implications for the Algorithm	237
The SIMPLS Algorithm	238
More on VIPs	244
The Standardize X Option	246
Determining the Number of Factors	246
Cross validation – How JMP does it	246

Ground Rules

The discussion in this appendix will assume a working knowledge of matrix algebra. There are many excellent references.

Many authors distinguish between PLS1 models (where each **Y** is modeled separately) and PLS2 models (where the **Y**s are modeled jointly), and there are many variants of PLS algorithms in each group. See Andersson (2009) for some variants of PLS1 algorithms. In this appendix, we only address JMP software's specific implementation of the NIPALS and SIMPLS methods.

In the following section, the matrix **X** is $n \times m$ and the matrix **Y** is $n \times k$.

The Singular Value Decomposition of a Matrix

Definition

There are various conventions in the literature regarding how the singular value decomposition is expressed. We will present the convention used by JMP. (This is also the convention used in LAPACK.)

Any $m \times k$ matrix **M** can be written as $\mathbf{M} = \mathbf{U}\Lambda\mathbf{V'}$:

- $r = \min(m, k)$
- $\mathbf{I}_r$ is an $r \times r$ identity matrix
- **U** is an $m \times r$ semi-orthogonal matrix ($\mathbf{U'U} = \mathbf{I}_r$),
- **V** is a $k \times r$ semi-orthogonal matrix ($\mathbf{V'V} = \mathbf{I}_r$),
- $\Lambda = diag(\lambda_1, \lambda_2, ..., \lambda_r)$ is an $r \times r$ diagonal matrix where $\lambda_1 \geq \lambda_2 \geq ... \geq \lambda_r \geq 0$,
- The symbol " ' " denotes the transpose of a matrix.

This representation of a matrix is called its *singular value decomposition*. Singular values and singular vectors are defined as follows:

- The diagonal entries of Λ are called the *singular values* of **M**.
- The r columns of **U** are called the *left singular vectors*.
- The r columns of **V** are called the *right singular vectors*.

Relationship to Spectral Decomposition

The singular value decomposition and the spectral decomposition of a square matrix have a close relationship. Writing out the relevant equations, you can verify that:

- The left singular vectors of **M** are eigenvectors of **MM'** (up to multiplication by −1).
- The right singular vectors of **M** are eigenvectors of **M'M** (up to multiplication by −1).
- The squares of the nonzero singular values of **M** are the nonzero eigenvalues of **M'M** (and **MM'**).

Other Useful Facts

Fact 1. **M** and **M'** have the same singular values.

Fact 2. Consider an $n \times m$ matrix **X**. Let w_1 denote the eigenvector of $A = X'X$ corresponding to the largest eigenvalue, λ_1. Then it follows from the spectral decomposition and the theory of quadratic forms that

$$\lambda_1 = (Xw_1)'(Xw_1) = \max_{\|f\|=1}\left[(Xf)'(Xf)\right]$$

Fact 3. Suppose that the $n \times m$ matrix **X** is centered. Then, since

$$Var(Xf) = (Xf)'(Xf)/(n-1)$$

it follows from Fact 2 that the largest eigenvalue of $X'X$ equals the maximum amount of variance explained by any norm-one linear combination of the columns of **X**. Also, that maximum variance is achieved when the linear combination is defined by the eigenvector of $X'X$ corresponding to the largest eigenvalue.

Principal Components Regression

Suppose that we want to use the $n \times m$ matrix **X** of predictors to predict the $n \times k$ matrix **Y** of response variables. Principal components regression uses Principal Components Analysis (PCA) to define factors that explain the variation in **X**. It is assumed that the predictors in **X** are at least centered. PCA proceeds by writing **X** in terms of its singular value decomposition, as described earlier:

$$X = U\Lambda V'$$

The squares of the nonzero singular values in Λ are the nonzero eigenvalues of $\mathbf{X'X}$.

The singular values are arranged in decreasing order and their corresponding singular vectors are placed in this order as well. As we have seen, the largest eigenvalue of $\mathbf{X'X}$ is associated with an eigenvector, $\mathbf{w}_1$, with the property that $\mathbf{Xw}_1$ has maximum variance among all norm one linear combinations of the columns of $\mathbf{X}$. The second largest eigenvalue gives the maximum variance among all linear combinations orthogonal to the first, and is defined by the second eigenvector. This continues for subsequent eigenvalues.

Now, recall that the right singular vectors are the eigenvectors of $\mathbf{X'X}$. In PCA, the right singular vectors in $\mathbf{V}$ are called the *factor loadings*. They define the directions of maximum variance. The vectors in $\mathbf{XV}$ are the *score vectors*, more commonly called *principal components* in the PCA literature. These are the projections of $\mathbf{X}$ onto the directions of maximum variance. (We note that the JMP PCA algorithm differs slightly from this description. For principal components on correlations, it scales the eigenvalues so that they sum to the number of columns of $\mathbf{X}$.)

In principal components regression, sufficiently many score vectors are retained and these are used to predict $\mathbf{Y}$. Because the score vectors are orthogonal, there are no issues with multicollinearity. However, there is no assurance that the subset of score vectors selected will be optimal in the sense of predicting $\mathbf{Y}$. The PCA scores are constructed to optimize accounting for variation in $\mathbf{X}$. Their relevance to predicting $\mathbf{Y}$ is not considered.

The Idea behind PLS Algorithms

PLS, on the other hand, attempts to construct factors from the $\mathbf{X}$ matrix that are relevant to predicting $\mathbf{Y}$. It does this by finding factors in the $\mathbf{X}$ space that maximize the covariance between $\mathbf{X}$ and $\mathbf{Y}$. These factors are then used as predictors for $\mathbf{Y}$. In this sense, PLS expands on PCA. Given that PLS factors are determined based on their relevance to $\mathbf{Y}$, usually fewer PLS factors than PCA factors are required to obtain a given level of predictive accuracy.

The matrix $\mathbf{X}$ can be fully decomposed as $\mathbf{X} = \mathbf{TP'}$, where $\mathbf{T}$ is a matrix whose columns are called the X scores, and where $\mathbf{P}$ is a matrix whose columns are called the X loadings. When $\mathbf{X}$ is fully decomposed, the number of columns in $\mathbf{T}$ equals the rank of $\mathbf{X}$. The matrix $\mathbf{Y}$ is modeled using linear regression on the X scores. In practice, because the goal is to model $\mathbf{X}$ and $\mathbf{Y}$ with a small number of factors, the matrix $\mathbf{X}$ is never fully decomposed. (We note that the scaling or normalization of score vectors is not standard among algorithms.)

As we have seen, the PLS algorithms extract factors in stages. The first stage is based either on the matrices **X** and **Y** (NIPALS) or on their covariance matrix **S** (SIMPLS). The next stage is based on matrices that are adjusted for the effects of extracting the first factor. We call this process *deflation*. Given that the *i*th factor has been extracted, the *i+1*st factor is extracted after deflating for the *i*th factor.

NIPALS

Applications of the NIPALS algorithm typically assume that the columns of the matrices **X** and **Y** have been both centered and scaled, although this is not required and in some cases, might not be desirable. However, to simplify the discussion in what follows, we will assume that both **X** and **Y** have been centered and scaled.

The NIPALS Algorithm

We will describe the JMP implementation of the algorithm. This is the standard implementation with the exception of normalizations, which vary among algorithms, as mentioned earlier. However, these normalizations do not affect predicted values.

Notation

We assume that **X** is $n \times m$ and **Y** is $n \times k$. Denote centered and scaled matrices corresponding to **X** and **Y** by $\mathbf{X}_{cs}$ and $\mathbf{Y}_{cs}$ (where "cs" stands for "centered and scaled"). That is, for any column of values in $\mathbf{X}_{cs}$ or $\mathbf{Y}_{cs}$, the mean is 0 and the standard deviation is 1.

All vectors and matrices are given in boldface, and vectors represent column vectors:

a

> This is the number of iterations of the algorithm, or equivalently, the number of factors extracted. The maximum number of factors is the rank of $\mathbf{X}_{cs}$: $a \leq rank(\mathbf{X}_{cs})$. As we have seen, the number of factors is often determined using cross validation.

$\mathbf{E}_i$, $\mathbf{F}_i$

> These represent the deflated matrices at each iteration of the algorithm. At the first step, $\mathbf{E}_1 = \mathbf{X}_{cs}$ and $\mathbf{F}_1 = \mathbf{Y}_{cs}$.

$\mathbf{w}_i$

The ith vector ($m \times 1$) of X weights.

$\mathbf{t}_i$

The ith vector ($n \times 1$) of X scores.

$\mathbf{c}_i$

The ith vector ($k \times 1$) of Y weights, also called Y loadings.

$\mathbf{u}_i$

The ith vector ($n \times 1$) of Y scores.

$\mathbf{p}_i$

The ith vector ($m \times 1$) of X loadings. The vector $\mathbf{p}_i$ contains normalized coefficients for a simple linear regression of the columns of $\mathbf{E}_i$ on the score vector $\mathbf{t}_i$. The larger in absolute value the regression coefficient in $\mathbf{p}_i$, the stronger the relationship of the corresponding predictor in $\mathbf{E}_i$ with the ith factor.

b_i

The regression coefficient for the regression of $\mathbf{u}_i$ on $\mathbf{t}_i$, namely, the regression of the Y scores on the X scores. This is thought of as a regression for the inner relation of the two data sets expressed in terms of their respective latent factors.

The Algorithm

The following algorithm is repeated until a factors have been extracted, or until the rank of $\mathbf{E}_{i+1}'\mathbf{F}_{i+1}$ is 0. In Steps 10 and 11, the current predicted values for $\mathbf{E}_i$ and $\mathbf{F}_i$ are calculated. These are subtracted from the current $\mathbf{E}_i$ and $\mathbf{F}_i$ matrices in Steps 12 and 13. The new matrices, $\mathbf{E}_{i+1}$ and $\mathbf{F}_{i+1}$, are residual matrices, obtained through the process of deflation.

At the ith iteration, the following steps are conducted:

1. Obtain the singular value decomposition of $\mathbf{E}_i'\mathbf{F}_i$.

2. Define $\mathbf{w}_i^0$ to be the first left singular vector of $\mathbf{E}_i'\mathbf{F}_i$.

3. Define $\mathbf{t}_i^0 = \mathbf{E}_i \mathbf{w}_i^0$.

4. Define $\mathbf{c}_i$ to be the first right singular vector of $\mathbf{E}_i'\mathbf{F}_i$.

5. Define $\mathbf{u}_i = \mathbf{F}_i \mathbf{c}_i$.

6. Define $\mathbf{p}^0_i = \mathbf{E}_i ' \mathbf{t}^0_i / (\mathbf{t}^0_i ' \mathbf{t}^0_i)$. Note that $\mathbf{p}^0_i$ contains regression coefficients for a regression of $\mathbf{E}_i$ on $\mathbf{t}^0_i$.

7. Define $\mathbf{p}_i = \mathbf{p}^0_i / \sqrt{\mathbf{p}^0_i ' \mathbf{p}^0_i}$.

8. Scale $\mathbf{t}^0_i$ and $\mathbf{w}^0_i$:

$$\mathbf{t}_i = \mathbf{t}^0_i \sqrt{\mathbf{p}^0_i ' \mathbf{p}^0_i}$$
$$\mathbf{w}_i = \mathbf{w}^0_i \sqrt{\mathbf{p}^0_i ' \mathbf{p}^0_i}$$

This scaling ensures that $\mathbf{p}_i$ contains regression coefficients for a regression of $\mathbf{E}_i$ on $\mathbf{t}_i$, so that $\mathbf{p}_i = \mathbf{E}_i ' \mathbf{t}_i / (\mathbf{t}_i ' \mathbf{t}_i)$. The vector $\mathbf{w}^0_i$ is adjusted accordingly, so that $\mathbf{t}_i = \mathbf{E}_i \mathbf{w}_i$.

9. Define $b_i = \mathbf{u}_i ' \mathbf{t}_i / (\mathbf{t}_i ' \mathbf{t}_i)$.

10. Compute the matrix $\mathbf{t}_i \mathbf{p}_i '$. This matrix contains predictions for the values in the matrix $\mathbf{E}_i$, based on the factor scores $\mathbf{t}_i$.

11. Compute the matrix $b_i \mathbf{t}_i \mathbf{c}_i '$. This matrix contains predictions for the values in the matrix $\mathbf{F}_i$, based on the factor scores $\mathbf{t}_i$. By way of intuition: for each response, the score vector $\mathbf{t}_i$ is multiplied by the appropriate Y weight; then the resulting matrix is multiplied by the regression coefficient b_i, which relates the Y scores to the X scores. A technical argument supporting the assertion is provided in the following section.

12. $\mathbf{E}_{i+1} = \mathbf{E}_i - \mathbf{t}_i \mathbf{p}_i '$.

13. $\mathbf{F}_{i+1} = \mathbf{F}_i - b_i \mathbf{t}_i \mathbf{c}_i '$.

14. Go back to Step 1 (using $\mathbf{E}_{i+1}$ and $\mathbf{F}_{i+1}$).

The vectors $\mathbf{w}_i, \mathbf{t}_i, \mathbf{p}_i, \mathbf{c}_i,$ and $\mathbf{u}_i$ and the scalars b_i, are stored in the matrices $\mathbf{W}, \mathbf{T}, \mathbf{P}, \mathbf{C}, \mathbf{U},$ and Δ_b:

$$\begin{aligned} \mathbf{W} &= (\mathbf{w}_1, \mathbf{w}_2, ..., \mathbf{w}_a) \\ \mathbf{T} &= (\mathbf{t}_1, \mathbf{t}_2, ..., \mathbf{t}_a) \\ \mathbf{P} &= (\mathbf{p}_1, \mathbf{p}_2, ..., \mathbf{p}_a) \\ \mathbf{C} &= (\mathbf{c}_1, \mathbf{c}_2, ..., \mathbf{c}_a) \\ \mathbf{U} &= (\mathbf{u}_1, \mathbf{u}_2, ..., \mathbf{u}_a) \\ \Delta_b &= diag(b_1, b_2, ..., b_a) \end{aligned}$$

Here, *diag* represents a diagonal matrix with the specified entries on the diagonal.

Computational Results

The E and F Models

For each extracted factor $\mathbf{t}_i$, predictive models for both $\mathbf{E}_i$ and $\mathbf{F}_i$ can be constructed by regressing $\mathbf{E}_i$ and $\mathbf{F}_i$ on $\mathbf{t}_i$:

$$\begin{aligned} \hat{\mathbf{E}}_i &= \mathbf{t}_i(\mathbf{t}_i'\mathbf{E}_i)/(\mathbf{t}_i'\mathbf{t}_i) = \mathbf{t}_i \mathbf{p}_i' \\ \hat{\mathbf{F}}_i &= \mathbf{t}_i(\mathbf{t}_i'\mathbf{F}_i)/(\mathbf{t}_i'\mathbf{t}_i) = b_i \mathbf{t}_i \mathbf{c}_i' \end{aligned}$$

In Proposition 2 below, we show that

$$\mathbf{t}_i(\mathbf{t}_i'\mathbf{F}_i)/(\mathbf{t}_i'\mathbf{t}_i) = b_i \mathbf{t}_i \mathbf{c}_i'$$

The scalar b_i is the regression coefficient for the regression of $\mathbf{u}_i$ on $\mathbf{t}_i$, which is the regression of the *i*th Y scores on the *i*th X scores. This is thought of as a regression for the inner relation of the two data sets defined by their respective latent factors. The predicted responses $b_i \mathbf{t}_i$ are assigned weights by the entries of $\mathbf{c}_i$, the right singular vector at step *i*.

It follows that the matrices $\mathbf{E}_{i+1}$ and $\mathbf{F}_{i+1}$ contain the residuals for the fits based on the *i*th extracted factor.

The Models for X and Y

The predicted values for each latent factor are summed to provide models for **X** and **Y**:

$$\hat{\mathbf{X}} = \sum_{i=1}^{a} \mathbf{t}_i \mathbf{p}_i' = \mathbf{TP}'$$

$$\hat{Y} = \sum_{i=1}^{a} b_i \mathbf{t}_i \mathbf{c}_i' = \mathbf{T}\Delta_b \mathbf{C}'$$

Using Proposition 3, which states that $\mathbf{T} = \mathbf{X}_{cs}\mathbf{W}(\mathbf{P}'\mathbf{W})^{-1}$, we can write

$$\hat{Y} = \mathbf{X}_{cs}\mathbf{W}(\mathbf{P}'\mathbf{W})^{-1}\Delta_b \mathbf{C}' = \mathbf{X}_{cs}\mathbf{B}$$

where

$$\mathbf{B} = \mathbf{W}(\mathbf{P}'\mathbf{W})^{-1}\Delta_b \mathbf{C}'$$

This gives the predicted values in terms of the centered and scaled predictors $\mathbf{X}_{cs}$, and can be adjusted to give the predicted values in terms of the untransformed predictors $\mathbf{X}$.

Distances to the X and Y Models

For each observation, distances to the X and Y models are computed in terms of the raw values. Consider the Y model. For a given observation, the difference between the predicted value and the observed value is computed. This is done for each column in $\mathbf{Y}$. These residuals are squared and divided by the variance of the observed values in the corresponding column of $\mathbf{Y}$. For each observation, these k values are summed. The square root of the sum is the distance to the Y model for that observation. The calculation for distances to the X model is similar.

Sums of Squares for Y

The sum of squares contribution for the fth factor to the Y model is defined as

$$SS(YModel)_f = Sum(Diag[(b_f \mathbf{t}_f \mathbf{c}_f')'(b_f \mathbf{t}_f \mathbf{c}_f')])$$

Loosely speaking, we can think of this sum of squares as reflecting the amount of variation in $\mathbf{Y}_{cs}$ explained by the fth factor. Note that $b_f \mathbf{t}_f$ is the vector of values predicted by the regression of $\mathbf{u}_f$ on $\mathbf{t}_f$. The entries of $b_f \mathbf{t}_f$ are weighted by the entries of $\mathbf{c}_f$, the right singular vector at step f, which contains the Y weights.

Define the total sum of squares for $\mathbf{Y}_{cs}$ as

$$SSY = Sum(Diag[\mathbf{Y}_{cs}'\mathbf{Y}_{cs}])$$
$$= \sum_{j=1}^{k}\sum_{i=1}^{n} y_{ij}^2$$

where $\mathbf{Y}_{cs} = (y_{ij})$.

The *Percent Variation Explained for Y Responses* for factor *f* is given by

$$\frac{SS(YModel)_f}{SSY}$$

Sums of Squares for X

Similarly, a sum of squares for the contribution of factor *f* to the X model is defined as

$$SS(XModel)_f = Sum(Diag[(\mathbf{t}_f \mathbf{p}_f')'(\mathbf{t}_f \mathbf{p}_f')])$$

The total sum of squares for the X Model is

$$SSX = Sum(Diag[\mathbf{X}_{cs}' \mathbf{X}_{cs}])$$
$$= \sum_{j=1}^{m} \sum_{i=1}^{n} x_{ij}^2$$

where $\mathbf{X}_{cs} = (x_{ij})$.

The *Percent Variation Explained for X Effects* for factor *f* is given by

$$\frac{SS(XModel)_f}{SSX}$$

The VIPs

The VIP values are calculated based on the model that is fit, which depends on the number of latent factors. Suppose that *a* factors are fit. Define

$$SS(YModel) = \sum_{f=1}^{a} SS(YModel)_f$$

The VIP for the *i*th predictor is defined as

$$VIP_i = \sqrt{m \sum_{f=1}^{a} \frac{w_{fi}^2 SS(YModel)_f}{\mathbf{w}_f' \mathbf{w}_f SS(YModel)}}$$

The size of the VIP for a predictor is driven by the product of its squared weights and the factor contributions to explaining variation in the responses. We can think of a predictor's VIP as reflecting its influence on the prediction of **Y** based on its role in determining the latent factor structure of the model. Note that the weights employed in defining the VIP for a predictor relate to the residuals for the predictor in the residual regressions.

An easy calculation shows that the sum of the squares of the VIPs over all predictors is $\sum_{i=1}^{m} VIP_i^2 = m$. It follows that the mean value for a predictor's squared VIP is one. This fact underlies the thinking that predictors with VIPs less than 0.8, or even 1.0, might not be influential for the model.

An alternative definition for VIPs is based on transforming the weights so that they apply to the original predictors $\mathbf{X}_{cs}$, rather than to the residuals $\mathbf{E}_i$. This approach uses the relationship derived in Proposition 3 found in the section "Transformation for Weights":

$$\mathbf{T} = \mathbf{X}_{cs}\mathbf{W}(\mathbf{P'W})^{-1}$$

The matrix $\mathbf{W}^* = \mathbf{W}(\mathbf{P'W})^{-1}$ can be considered a weight matrix that gives the factor scores in terms of the original predictors, $\mathbf{X}_{cs}$. The matrix $\mathbf{W}^*$ can be normalized and the entries used as weights, in analogy with the definition of VIP given earlier, to obtain VIP values that we refer to as VIP*. This definition is alluded to in (Wold et al., 1993, pp. 547–548). We study the VIP* values in Appendix 2.

Properties of the NIPALS Algorithm

This section lists properties of the matrices and vectors involved in the NIPALS algorithm. Although we prove some of these results in this section, for others, we will simply cite the source of the proof.

In the following, let $\mathbf{E}$ and $\mathbf{F}$, without subscripts, denote residual matrices at any iteration of the NIPALS algorithm. Denote by λ_1 the first (largest in absolute value) singular value of $\mathbf{E'F}$. Let $\mathbf{w}$ and $\mathbf{c}$ denote the corresponding X and Y weights. We will revert to using subscripts only when needed for clarity.

It is important to note that, in the NIPALS algorithm, the residual matrices $\mathbf{E}$ and $\mathbf{F}$ are used to compute each new factor. This means that the weights, scores, and loadings computed during each iteration of the algorithm relate to those residual matrices. By contrast, as we will see in the section "SIMPLS", the SIMPLS algorithm computes quantities for each factor that are based on repeatedly deflating $\mathbf{X'Y}$, the covariance matrix of $\mathbf{X}$ and $\mathbf{Y}$.

Properties of the X Weights, Scores, and Loadings

Proofs of the first three results below are presented in the paper by Hoskuldsson (1988, Properties 1–3), which goes into great detail about properties of the algorithm. Property 4 can be verified directly from the definition in the NIPALS algorithm.

1. The vectors $\mathbf{w}_i$ are mutually orthogonal for $i = 1,\ldots,a$.
2. The vectors $\mathbf{t}_i$ are mutually orthogonal for $i = 1,\ldots,a$.
3. For $i < j$, the vectors $\mathbf{w}_i$ are orthogonal to the vectors $\mathbf{p}_j$.
4. For any i, $\mathbf{p}_i'\mathbf{w}_i = 1$.

Maximization of Covariance

The Y weight, $\mathbf{c}$, is a right singular vector, so it has norm 1. In the JMP implementation of NIPALS, the X weight $\mathbf{w}$ is scaled so that $\mathbf{w} = \mathbf{w}^0\sqrt{\mathbf{p}^{0\prime}\mathbf{p}^0} = k\mathbf{w}^0$, where $\mathbf{w}^0$ is a left singular vector. It follows that $\mathbf{w}$ has norm k.

The algorithm ensures that, for each set of residual matrices, the weights $\mathbf{w}$ and $\mathbf{c}$ define linear combinations of the variables in $\mathbf{E}$ and $\mathbf{F}$ that maximize the covariance among all linear combinations defined by vectors with the same norms. Because the X and Y scores are given, respectively, by $\mathbf{t} = \mathbf{E}\mathbf{w}$ and $\mathbf{u} = \mathbf{F}\mathbf{c}$, it follows that the X and Y scores have maximum covariance at each iteration of the algorithm, among all linear combinations with the specified norms. This is a fundamental underpinning of the PLS methodology. This result is stated and verified in Proposition 1.

Proposition 1. Suppose that $\mathbf{X}$ and $\mathbf{Y}$ are centered. Then the vectors $\mathbf{w}^0$ and $\mathbf{c}$ define linear combinations of the columns of $\mathbf{E}$ and the columns of $\mathbf{F}$, respectively, that have maximum covariance among all norm-one vectors. In symbols:

$$Cov(\mathbf{E}\mathbf{w}^0, \mathbf{F}\mathbf{c}) = \max_{\|f\|=\|g\|=1} Cov(\mathbf{E}f, \mathbf{F}g)$$

Verification. The verification will be provided in steps.

Step 1. Suppose that $\mathbf{X}$ and $\mathbf{Y}$ are centered. Then it follows that all residual matrices $\mathbf{E}$ and $\mathbf{F}$ are centered as well. To see this, we use induction. We present the argument for $\mathbf{E}$; the argument for $\mathbf{F}$ is similar.

Suppose that $\mathbf{E}_i$ is centered for some i. Then:

$$\mathbf{E}_{i+1} = \mathbf{E}_i - \mathbf{t}_i\mathbf{p}_i'$$

$$\mathbf{t}_i = \mathbf{t}_i^0 \sqrt{\mathbf{p}_i^{0'}\mathbf{p}_i^0} = \mathbf{E}_i \mathbf{w}_i^0 \sqrt{\mathbf{p}_i^{0'}\mathbf{p}_i^0} \propto \mathbf{E}_i \mathbf{w}_i^0,$$

where the symbol "$\propto$" denotes proportionality. Given that $\mathbf{E}_i$ is centered, it follows that $\mathbf{t}_i$ and hence $\mathbf{E}_{i+1}$ are centered.

Step 2. For any vectors $\mathbf{f}$ and $\mathbf{g}$, $(n-1)Cov(\mathbf{Ef}, \mathbf{Fg}) = \mathbf{f'E'Fg}$. This follows from the definition of covariance and the fact that $\mathbf{E}$ and $\mathbf{F}$ are centered:

$$(n-1)Cov(\mathbf{Ef}, \mathbf{Fg}) = [(\mathbf{Ef} - \overline{\mathbf{Ef}})'(\mathbf{Fg} - \overline{\mathbf{Fg}})] = \mathbf{f'E'Fg}$$

where the bar over a vector indicates a vector whose elements are its mean.

Step 3. For the vectors $\mathbf{w}^0$ and $\mathbf{c}$,

$$Cov(\mathbf{Ew}^0, \mathbf{Fc}) = \lambda_1 / (n-1)$$

To see this, let $\mathbf{U\Lambda V'}$ denote the singular value decomposition of $\mathbf{E'F}$. Denote the first singular value by λ_1. Using the fact that $\mathbf{E}$ and $\mathbf{F}$ are centered and that $\mathbf{U}$ and $\mathbf{V}$ are semi-orthogonal,

$$(n-1)Cov(\mathbf{Ew}^0, \mathbf{Fc}) = (\mathbf{w}^0)'\mathbf{E'Fc} = (\mathbf{w}^0)'\mathbf{U\Lambda V'c} = \lambda_1$$

Step 4. Suppose that $\|\mathbf{f}\| = \|\mathbf{g}\| = 1$. Then $|Cov(\mathbf{Ef}, \mathbf{Fg})| \leq Cov(\mathbf{Ew}^0, \mathbf{Fc})$. The verification proceeds as follows:

$$\begin{aligned}
\left[(n-1)Cov(\mathbf{Ef}, \mathbf{Fg})\right]^2 &= \left[\mathbf{f'E'Fg}\right]^2 \\
&= \left[(\mathbf{F'Ef})'\mathbf{g}\right]^2 \\
&\leq \|\mathbf{F'Ef}\|^2 \|\mathbf{g}\|^2 \\
&= \|\mathbf{F'Ef}\|^2 \\
&= \mathbf{f'(F'E)'(F'E)f} \\
&\leq \lambda_1^2 \\
&= \left[(n-1)Cov(\mathbf{Ew}^0, \mathbf{Fc})\right]^2
\end{aligned}$$

Here, the first inequality follows from the Cauchy-Schwarz Inequality. The second inequality follows from Fact 2 in the section "The Singular Value Decomposition of a Matrix" and the fact that λ_1^2 is the maximum eigenvalue of $(\mathbf{F'E})'(\mathbf{F'E}) = (\mathbf{E'F})(\mathbf{E'F})'$.

Bias toward X Directions with High Variance

It follows from Proposition 1 that the weight $\mathbf{w}^0$ satisfies

$$Corr^2(\mathbf{Ew}^0, \mathbf{Fc})Var(\mathbf{Ew}^0) = \max_{\|f\|=1}\left[Corr^2(\mathbf{Ef}, \mathbf{Fc})Var(\mathbf{Ef})\right]$$

To see this, note that $\mathbf{c}$ has norm one so that

$$Cov(\mathbf{Ew}^0, \mathbf{Fc}) = \max_{\|f\|=1} Cov(\mathbf{Ef}, \mathbf{Fc})$$

Now write the covariances in terms of correlations.

This result shows that the X weights, and hence the X scores, attempt to maximize both correlation with the **Y** structure and variance in the **X** structure. As a result, the coefficients for the PLS model are biased away from **X** directions with small variance.

Regression Coefficient for Inner Relation

At each iteration of the NIPALS algorithm, the inner regression relationship consists of regressing the Y scores, $\mathbf{u}$, on the X scores, $\mathbf{t}$. In the section "The E and F Models," we expressed the predicted **F** matrix in terms of the inner regression relation's coefficient, b. Proposition 2 verifies this representation.

Proposition 2. The model for **F**, obtained by regressing on $\mathbf{t}$, can be expressed as

$$\hat{\mathbf{F}} = \mathbf{t}(\mathbf{t'F})/(\mathbf{t't}) = b\mathbf{tc'},$$

where b is the regression coefficient for the inner relation.

Verification. The first equality follows from the definition of regression. We will verify the second equality. Let $\mathbf{U\Lambda V'}$ denote the singular value decomposition of $\mathbf{E'F}$. For any iteration of the algorithm, denote the first singular value by λ_1, the first X score by $\mathbf{t}$, the first Y score by $\mathbf{u}$, and the first right singular vector by $\mathbf{c}$. With this notation, the regression coefficient for the inner relation, b, can be written

$$b = \mathbf{u't}/(\mathbf{t't})$$

We will proceed in steps. The following steps hold for any iteration of the algorithm.

Step 1. $\mathbf{F't} = k\lambda_1\mathbf{c}$. This can be verified using the singular value decomposition and the property that the singular vectors in **U** and **V** are orthonormal:

$$\mathbf{F't} = \mathbf{F'Ew} = k(\mathbf{U\Lambda V'})'\mathbf{w}^0 = k\mathbf{V\Lambda U'w}^0 = k\lambda_1\mathbf{c}$$

Step 2. $\mathbf{u't} = k\lambda_1$. From Step 3 in the proof of Proposition 1, $Cov(\mathbf{Ew}^0, \mathbf{Fc}) = \lambda_1/(n-1)$. It follows that $\mathbf{u't} = \mathbf{t'u} = (n-1)Cov(\mathbf{t},\mathbf{u}) = (n-1)Cov(k\mathbf{Ew}^0, \mathbf{Fc}) = k\lambda_1$.

Step 3. $\mathbf{t(t'F)} = (\mathbf{u't})\mathbf{tc'}$. Using the results of Steps 1 and 2,

$$\begin{aligned}\mathbf{t(t'F)} &= \mathbf{t(F't)'} \\ &= \mathbf{t}(k\lambda_1\mathbf{c})' \\ &= k\lambda_1\mathbf{tc'} \\ &= (\mathbf{u't})\mathbf{tc'}\end{aligned}$$

This shows that $\mathbf{t(t'F)}/(\mathbf{t't}) = [\mathbf{u't}/(\mathbf{t't})]\mathbf{tc'} = b\mathbf{tc'}$.

The Loadings as Measures of Correlation with the Factors

Suppose that $\mathbf{X}$ and $\mathbf{Y}$ are both centered and scaled. Consider the ith iteration. The ith vector of X loadings, $\mathbf{p}_i$, is defined by the algorithm to have the property that

$$\mathbf{p}_i \propto \mathbf{E}_i'\mathbf{t}_i$$

But

$$\mathbf{E}_i = \mathbf{X}_{cs} - \sum_{k=1}^{i-1} \mathbf{t}_k \mathbf{p}_k'$$

It follows that

$$\mathbf{p}_i \propto \mathbf{E}_i'\mathbf{t}_i = \mathbf{X}_{cs}'\mathbf{t}_i - \left(\sum_{k=1}^{i-1}\mathbf{t}_k\mathbf{p}_k'\right)'\mathbf{t}_i = \mathbf{X}_{cs}'\mathbf{t}_i - \sum_{k=1}^{i-1}\mathbf{p}_k\mathbf{t}_k'\mathbf{t}_i = \mathbf{X}_{cs}'\mathbf{t}_i$$

since the $\mathbf{t}_i$ are orthogonal. Because $\mathbf{X}_{cs}$ is centered and scaled, we see that $\mathbf{p}_i$ is proportional to the correlations between the centered and scaled predictors and the X score $\mathbf{t}_i$. (Recall that JMP scales all loading vectors to have norm 1.)

The Y loadings have a similar property. It is shown in the demonstration of Proposition 2 that

$$\mathbf{c}_i \propto \mathbf{F}_i'\mathbf{t}_i$$

In a fashion similar to our derivation for $\mathbf{p}_i$, we can show that

$$\mathbf{c}_i \propto \mathbf{F}_i'\mathbf{t}_i = \mathbf{Y}_{cs}'\mathbf{t}_i$$

It follows that the elements of $\mathbf{c}_i$ are proportional to the correlations of the centered and scaled responses with t_i.

Transformation for Weights

The weights computed by NIPALS are used to define the score vectors. But these weights are derived from the residual matrices. The next proposition shows how to write the score matrix in terms of the original variables in $\mathbf{X}$.

Proposition 3. $\mathbf{T} = \mathbf{X}_{cs}\mathbf{W}(\mathbf{P'W})^{-1}$

Verification. In our notation, a iterations are conducted to obtain the matrices $\mathbf{T}$, $\mathbf{W}$, and $\mathbf{P}$. Had $rank(\mathbf{X})$ iterations been conducted, we would be able to write

$$\begin{aligned}
\mathbf{X}_{cs} &= \mathbf{E}_1 \\
&= \mathbf{E}_2 + \mathbf{t}_1\mathbf{p}_1' \\
&= \mathbf{E}_3 + \mathbf{t}_2\mathbf{p}_2' + \mathbf{t}_1\mathbf{p}_1' \\
&= \ldots \\
&= \sum_{i=1}^{rank(X)} \mathbf{t}_i\mathbf{p}_i' \\
&= \sum_{i=1}^{a} \mathbf{t}_i\mathbf{p}_i' + \sum_{i=a+1}^{rank(X)} \mathbf{t}_i\mathbf{p}_i'
\end{aligned}$$

Then,

$$\begin{aligned}
\mathbf{X}_{cs}\mathbf{W} &= \left(\sum_{i=1}^{a} \mathbf{t}_i\mathbf{p}_i'\right)\mathbf{W} + \left(\sum_{i=a+1}^{rank(X)} \mathbf{t}_i\mathbf{p}_i'\right)\mathbf{W} \\
&= \mathbf{TP'W} + \left(\sum_{i=a+1}^{rank(X)} \mathbf{t}_i\mathbf{p}_i'\right)\mathbf{W} \\
&= \mathbf{TP'W} + \left(\sum_{i=a+1}^{rank(X)} \mathbf{t}_i\mathbf{p}_i'\mathbf{w}_1, \sum_{i=a+1}^{rank(X)} \mathbf{t}_i\mathbf{p}_i'\mathbf{w}_2, \ldots, \sum_{i=a+1}^{rank(X)} \mathbf{t}_i\mathbf{p}_i'\mathbf{w}_a\right) \\
&= \mathbf{TP'W}
\end{aligned}$$

The last equality holds because of the third property in the section "Properties of the X Weights, Scores, and Loadings," which states, "For $i < j$, the vectors $\mathbf{w}_i$ are orthogonal to the vectors $\mathbf{p}_j$."

That **P'W** is invertible follows from the fact that it is an $a \times a$ upper-triangular matrix with one's on the main diagonal (Properties 3 and 4 in "Properties of the X Weights, Scores, and Loadings").

SIMPLS

Applications of the SIMPLS algorithm also typically assume that the matrices **X** and **Y** have been centered and scaled. We will present the algorithm from this perspective. To emphasize the centering and scaling, we continue to write the centered and scaled matrices as $\mathbf{X}_{cs}$ and $\mathbf{Y}_{cs}$. The JMP implementation of the algorithm is the standard implementation (de Jong 1993), but with additional normalizations.

Optimization Criterion

Before describing the algorithm, we present some background. De Jong's goal in developing SIMPLS was to first specify an optimization criterion, and then develop an algorithm that fulfilled that criterion. This is in contrast to NIPALS, which is a methodology defined by an algorithm.

The idea behind SIMPLS is to find a predictive linear model for **Y** by extracting successive orthogonal factors from **X**. In SIMPLS, each factor is determined in a way that maximizes the covariance with corresponding linear combinations of the columns of **Y**. Specifically, the scores are defined as $\mathbf{t}_i \propto \mathbf{X}_{cs}\mathbf{w}_i$, where the vectors $\mathbf{w}_i$ and $\mathbf{c}_i$ satisfy the following:

- $\text{Cov}(\mathbf{X}_{cs}\mathbf{w}_i, \mathbf{Y}_{cs}\mathbf{c}_i)$ maximizes $\text{Cov}(\mathbf{X}_{cs}\mathbf{f}, \mathbf{Y}_{cs}\mathbf{g})$ over all vectors **f** and **g** of length one, namely, where $\mathbf{f'f} = \mathbf{g'g} = 1$;
- The X scores are orthogonal; that is, for $i \neq j$, we require that $\mathbf{t}_i'\mathbf{t}_j = 0$.

Note that, in NIPALS, the covariance is maximized for components defined on the residual matrices. In contrast, the maximization in SIMPLS applies directly to $\mathbf{X}_{cs}$ and $\mathbf{Y}_{cs}$.

These two criteria define de Jong's objective and drive the details of the algorithm. We outline some of these details in the remainder of this section.

Implications for the Algorithm

For each score $\mathbf{t}_i$, a corresponding loading vector is defined as $\mathbf{p}_i = \mathbf{X}_{cs}'\mathbf{t}_i$. The requirement that the X scores be orthogonal implies that any weight vector is orthogonal to all preceding loading vectors. That is, for $i < j$, the fact that $\mathbf{p}_i'\mathbf{w}_j = 0$ follows from

$$\mathbf{p}_i'\mathbf{w}_j = (\mathbf{X}_{cs}'\mathbf{t}_i)'\mathbf{w}_j = \mathbf{t}_i'\mathbf{X}_{cs}\mathbf{w}_j \propto \mathbf{t}_i'\mathbf{t}_j = 0$$

Note that, in NIPALS, the matrix $\mathbf{P'W}$ is upper triangular. In SIMPLS, it is diagonal.

For $k > 1$, denote the matrix of projection vectors $\mathbf{p}_1, \mathbf{p}_2, ..., \mathbf{p}_{k-1}$ by $\mathbf{P}_{k-1}$. Then we require that $\mathbf{w}_k$ be orthogonal to the column space of $\mathbf{P}_{k-1}$. The projection matrix associated with $\mathbf{P}_{k-1}$ is $\mathbf{P}_{k-1}(\mathbf{P}_{k-1}'\mathbf{P}_{k-1})^{-1}\mathbf{P}_{k-1}'$. The matrix that projects onto the orthogonal subspace is

$$\mathbf{P}_{k-1}^\perp = \mathbf{I}_m - \mathbf{P}_{k-1}(\mathbf{P}_{k-1}'\mathbf{P}_{k-1})^{-1}\mathbf{P}_{k-1}'$$

It follows that $\mathbf{w}_k = \mathbf{P}_{k-1}^\perp \mathbf{w}_k$.

Define $\mathbf{S}_1 = \mathbf{X}_{cs}'\mathbf{Y}_{cs}$ ($\mathbf{S}_1$ is $m \times k$). Then for any vectors $\mathbf{w}_k$ and $\mathbf{c}_k$:

$$\begin{aligned}(n-1)Cov(\mathbf{X}_{cs}\mathbf{w}_k, \mathbf{Y}_{cs}\mathbf{c}_k) &= \mathbf{w}_k'\mathbf{X}_{cs}'\mathbf{Y}_{cs}\mathbf{c}_k \\ &= \mathbf{w}_k'\mathbf{S}_1\mathbf{c}_k \\ &= \mathbf{w}_k'(\mathbf{P}_{k-1} + \mathbf{P}_{k-1}^\perp)\mathbf{S}_1\mathbf{c}_k \\ &= \mathbf{w}_k'\mathbf{P}_{k-1}\mathbf{S}_1\mathbf{c}_k + \mathbf{w}_k'\mathbf{P}_{k-1}^\perp\mathbf{S}_1\mathbf{c}_k \\ &= 0 + \mathbf{w}_k'\mathbf{P}_{k-1}^\perp\mathbf{S}_1\mathbf{c}_k \\ &= \mathbf{w}_k'\mathbf{P}_{k-1}^\perp\mathbf{S}_1\mathbf{c}_k\end{aligned}$$

The requirement that $Cov(\mathbf{X}_{cs}\mathbf{f}, \mathbf{Y}_{cs}\mathbf{g})$ be maximized over all vectors $\mathbf{f}$ and $\mathbf{g}$ of length one implies that $\mathbf{w}_k$ and $\mathbf{c}_k$ are given by the first pair of singular vectors from the SVD of $\mathbf{P}_{k-1}^\perp \mathbf{S}_1$.

For $k > 1$, define $\mathbf{S}_k = \mathbf{P}_{k-1}^\perp \mathbf{S}_1$. Then the weight vectors that maximize the desired covariance are the first left and right singular vectors of $\mathbf{S}_k$.

To simplify the algorithm, the column space of $\mathbf{P}_{k-1}$ is represented by an orthonormal basis. Specifically, a Gram-Schmidt process is used to obtain an orthonormal basis. These basis vectors are denoted by $\mathbf{v}_1, \mathbf{v}_2, ..., \mathbf{v}_{k-1}$.

The SIMPLS Algorithm

Notation

As before, we assume that the $n \times m$ matrix $\mathbf{X}$ and the $n \times k$ matrix $\mathbf{Y}$ are centered and scaled, and we denote these matrices by $\mathbf{X}_{cs}$ and $\mathbf{Y}_{cs}$. That is, for any column of values in $\mathbf{X}_{cs}$ or $\mathbf{Y}_{cs}$, the mean is 0 and the standard deviation is 1.

All vectors and matrices are given in boldface, and vectors represent column vectors:

a

This is the number of iterations of the algorithm, or equivalently, the number of factors extracted. The maximum number of factors is the rank of $\mathbf{X}_{cs}$: $a \leq \text{rank}(\mathbf{X}_{cs})$.

$\mathbf{S}_i$

The deflated covariance matrix at each iteration of the algorithm. At the first step, $\mathbf{S}_1 = \mathbf{X}_{cs}'\mathbf{Y}_{cs}$

$\mathbf{w}_i$

The ith vector ($m \times 1$) of X weights

$\mathbf{t}_i$

The ith vector ($n \times 1$) of X scores

$\mathbf{c}_i$

The ith vector ($k \times 1$) of Y weights; also called Y loadings

$\mathbf{u}_i$

The ith vector ($n \times 1$) of Y scores

$\mathbf{p}_i$

The ith vector ($m \times 1$) of X loadings. The (column) vector $\mathbf{p}_i$ contains the coefficients for simple linear regressions of each of the columns of $\mathbf{X}_{cs}$ on the (length 1) score vector $\mathbf{t}_i$. The larger in absolute value the regression coefficient in $\mathbf{p}_i$, the stronger the relationship of the corresponding predictor in $\mathbf{X}_{cs}$ with the ith factor.

$\mathbf{v}_i$

The ith vector in the Gram-Schmidt orthonormal basis for $(\mathbf{p}_1, \mathbf{p}_2, ..., \mathbf{p}_i)$

$\mathbf{T}_i$

The matrix $(\mathbf{t}_1, \mathbf{t}_2, ..., \mathbf{t}_i)$

$\mathbf{V}_i$

The matrix $(\mathbf{v}_1, \mathbf{v}_2, ..., \mathbf{v}_i)$

The Algorithm

Define $S_1 = X_{cs}'Y_{cs}$. At the ith iteration, the following steps are conducted. Note that the weights and X scores are normalized using the X scores. This is done to simplify subsequent formulas. The steps are repeated until a factors have been extracted, or until the rank of S_{i+1} is 0.

1. Obtain the singular value decomposition of S_i.

2. Define w_i^0 to be the first left singular vector of S_i. Note that w_i^0 has length one.

3. Define $t_i^0 = X_{cs}w_i^0$.

4. Compute $norm(t_i^0) = \sqrt{t_i^0 {'} t_i^0}$.

5. Normalize t_i : $t_i = t_i^0 / norm(t_i^0)$. This normalizes the vector of X scores.

6. Normalize w_i : $w_i = w_i^0 / norm(t_i^0)$. This normalizes the weights in accordance with the scores.

7. Define c_i to be $Y_{cs}'t_i$. This is proportional to the right singular vector of S_i. More specifically, $c_i = \lambda_1 c_i^0 / \sqrt{t_i't_i}$, where λ_1 is the first singular value and c_i^0 is the first right singular vector of S_i.

8. Define $u_i^0 = Y_{cs}c_i$.

9. Define $p_i = X_{cs}'t_i$.

10. For all iterations other than the first, define $u_i = u_i^0 - T_{i-1}(T_{i-1}'u_i^0)$. This step constructs the u_i as transformed Y scores that are orthogonal to the preceding X scores. This transformation allows for easier interpretation and comparison to NIPALS and preserves the property that u_i and t_i have maximum covariance at each step.

11. Construct an orthonormal basis of vectors v_i for projection onto the orthogonal subspace. This enables one to compute S_{i+1} from S_i using only v_i :

    a) Set $v_1^0 = p_1$.

    b) For all iterations other than the first, set $v_i^0 = p_i - V_{i-1}(V_{i-1}'p_i)$.

    c) Normalize v_i^0 : $v_i = v_i^0 / \sqrt{v_i^0{'}v_i^0}$.

12. The deflated matrix S_{i+1} is computed as $S_{i+1} = S_i - v_i v_i' S_i$.

13. Go back to Step 1 (using S_{i+1}).

JMP Customizations

JMP applies a number of transformations to SIMPLS results in order to make them comparable to the NIPALS results:

1. The X weights and X scores are multiplied by the corresponding p-norms ($\sqrt{p_i' p_i}$).

2. The Y scores are divided by the norm of the Y loadings ($\sqrt{c_i' c_i}$).

3. The X and Y loadings are normalized.

The Models for X and Y

We continue with the notation established prior to the description of the JMP customizations. Define the matrices W, T, P, and C to contain their affiliated a columns.

The model for Y is obtained by regressing Y on T. Because the score vectors t_i are normalized, this regression equation is given by

$$\hat{Y} = TT'Y_{cs} = T(T'Y_{cs}) = TC' = X_{cs}WC' = X_{cs}B$$

where

$$B = WC'$$

Note that the weight matrix, W, applies directly to the predictor variables in X_{cs}. (This is in contrast to the situation in NIPALS, where the matrix of regression coefficients is $B = W(P'W)^{-1} \Delta_b C'$.)

As in NIPALS, the model for X is

$$\hat{X} = TT'X_{cs} = TP'$$

Distances to the X and Y Models

Distances to the X and Y models are computed as the square roots of the sums of squared scaled residuals. These are computed in terms of the raw data, rather than the centered and scaled values. (Details are given in the section "Computational Results" in "NIPALS".)

Sums of Squares for Y

The sum of squares contribution for the *f*th factor to the Y model is defined as

$$SS(YModel)_f = Sum(Diag[(\mathbf{t}_f\mathbf{c}_f')'(\mathbf{t}_f\mathbf{c}_f')])$$

We can think of $SS(YModel)_f$ as reflecting the amount of variation in $\mathbf{Y}_{cs}$ explained by the *f*th factor.

Define

$$SSY = Sum(Diag[\mathbf{Y}_{cs}'\mathbf{Y}_{cs}])$$
$$= \sum_{j=1}^{k}\sum_{i=1}^{n} y_{ij}^2$$

The *Percent Variation Explained for Y Responses* for factor *f* is given by

$$\frac{SS(YModel)_f}{SSY}$$

Sums of Squares for X

The sum of squares for the contribution of factor *f* to the X model is defined as

$$SS(XModel)_f = Sum(Diag[(\mathbf{t}_f\mathbf{p}_f')'(\mathbf{t}_f\mathbf{p}_f')])$$

The total sum of squares for the X Model is

$$SSX = Sum(Diag[\mathbf{X}_{cs}'\mathbf{X}_{cs}])$$
$$= \sum_{j=1}^{m}\sum_{i=1}^{n} x_{ij}^2$$

and the *Percent Variation Explained for X Effects* for factor *f* is given by

$$\frac{SS(XModel)_f}{SSX}$$

The Loadings as Measures of Correlation with the Factors

For the SIMPLS algorithm, the ith vector of X loadings, $\mathbf{p}_i$, is defined by

$$\mathbf{p}_i = \mathbf{X}_{cs}'\mathbf{t}_i$$

JMP then normalizes each $\mathbf{p}_i$. It follows that the elements of $\mathbf{p}_i$ are proportional to the correlations of the centered and scaled predictors with $\mathbf{t}_i$, which represents the ith factor. (Recall that JMP divides all loading vectors by their length so that they have norm 1.)

The Y loadings have a similar property. The Y loadings are defined by

$$\mathbf{c}_i = \mathbf{Y}_{cs}'\mathbf{t}_i$$

JMP normalizes the $\mathbf{c}_i$. It follows that the elements of $\mathbf{c}_i$ are proportional to the correlations of the centered and scaled responses with $\mathbf{t}_i$.

The VIPs

The VIP for the ith predictor is defined as

$$VIP_i = \sqrt{m \sum_{f=1}^{p} \frac{w_{fi}^2 SS(Model)_f}{\mathbf{w}_f'\mathbf{w}_f SS(Model)}}$$

Symbolically, this equation is identical to the one used to define VIPs for the NIPALS algorithm. However, the weights used in the two algorithms are defined differently.

In the case of SIMPLS, the weights satisfy $\mathbf{T} = \mathbf{X}_{cs}\mathbf{W}$. These SIMPLS weights relate directly to the original predictor values in $\mathbf{X}_{cs}$. In contrast, the weights used in defining VIPs in the case of NIPALS relate to the deflated matrices, namely, the residuals for the predictors in the residual regressions. In NIPALS, the weights are related to the original predictors through the relationship $\mathbf{T} = \mathbf{X}_{cs}\mathbf{W}(\mathbf{P}'\mathbf{W})^{-1}$.

As in NIPALS, it is easy to show that the sum of the squares of the SIMPLS VIPs over all predictors is $\sum_{i=1}^{m} VIP_i^2 = m$. It follows that the mean value for a predictor's squared VIP is one. One can extend the NIPALS guideline that predictors with VIPs less than 0.8, or even 1.0, might not be influential for the model. We explore these guidelines in a simulation study in Appendix 2.

It is important to note that the VIPs obtained using NIPALS and SIMPLS can be quite different. In particular, the numbers of predictors exceeding a threshold of 1.0 or 0.8 can differ substantially. This can occur even when the models given by the two approaches are very similar, as they often are. One must realize that the VIPs are of two different natures. We note in passing that the NIPALS VIP* values (which JMP 11 does not directly calculate) tend to be similar to the SIMPLS VIP values. We explore the three VIP types in a simulation study in Appendix 2.

More on VIPs

To better understand the NIPALS weights and their use in VIPs, we will look at a tiny example. The data table is called TinyDemoVIP.jmp, and you can open this by clicking on the correct link in the master journal. The table has four rows, five **X**s, and two **Y**s. The **Y**s are obtained by simulation, with Y1 a function of X1 only and Y2 a function or X2 only.

The script Fit Model Launch Window shows the model specification in Fit Model. The script Three Factor Models fits both NIPALS and SIMPLS models to the data. The fits, performed without validation, extract three factors.

The script Scores and Residuals computes the three X scores, and places them in the data table in columns called T1, T2, and T3. For each score, the associated $\mathbf{E}_i$ matrix is computed. Each matrix consists of five columns. These matrices are added as columns to the data table and are called First Residuals, Second Residuals, and Third Residuals. The first of these matrices is simply the centered and scaled **X** matrix.

The script also adds columns called T1 Calc, T2 Calc, and T3 Calc to the data table. The script also produces a new table called **Weights** containing three weights, W1, W2, and W3. These weight columns are used in the calculation of T1 Calc, T2 Calc, and T3 Calc. This enables you to verify that the scores are simply the linear combinations of the residual vectors multiplied by the weights.

Recall that the NIPALS VIPs are defined in terms of these weights. This simple example gives insight on how these weights are interpreted. They are the weights applied to the residuals in obtaining the X scores.

The script Table of WStar Values gives a table containing the $\mathbf{W}^*$ weights, namely, the weights $\mathbf{W}^* = \mathbf{W}(\mathbf{P'W})^{-1}$. These apply directly to the predictors in terms of obtaining the scores $\mathbf{T} = \mathbf{X}_{cs}\mathbf{W}(\mathbf{P'W})^{-1}$. To verify that $\mathbf{T} = \mathbf{X}_{cs}\mathbf{W}^*$, open a log window (**View > Log**) and run the script. The last line of code computes the product $\mathbf{X}_{cs}\mathbf{W}^*$.

The script VIP Comparison compares the VIP values obtained in JMP using NIPALS to the VIP* values obtained using the $\mathbf{W}^*$ weights (which are calculated directly by the script). The Three Factor Models script gives a report for a SIMPLS fit. Note that the values in the **Variable Importance Table** are extremely similar to the NIPALS VIP* values we obtain using the $\mathbf{W}^*$ weights.

Also note that, using the 0.8 or 1.0 cut-offs for VIPs can lead to different predictors being retained in a pruned model. You can explore differences by simulating new values for Y1 and Y2. Click on the **+** sign to the right of the column names in the **Columns** panel, click **Apply**, and rerun the scripts of interest.

Close TinyDemoVIP.jmp and any open reports generated by the scripts in it.

The script Compare_NIPALS_VIP_and_VIPStar.jsl gives additional insight into the differences between VIP and VIP* values from NIPALS fits. You can run the script by clicking on the correct link in the master journal. It generates sample data from an underlying model that you specify in the launch widow, performs a NIPALS fit, and then shows graphs of the VIP and VIP* values for each **X** term in the data, together with their differences. You can specify the number of simulations by setting the **Number of Repeats** (3 is the minimum number). A Graph Builder plot comparing VIP to VIP* is shown for each simulation. Accepting the defaults gives a report similar to that in Figure A1.1.

FigureA1.1: Comparing VIP and VIP* Values for Simulated Data

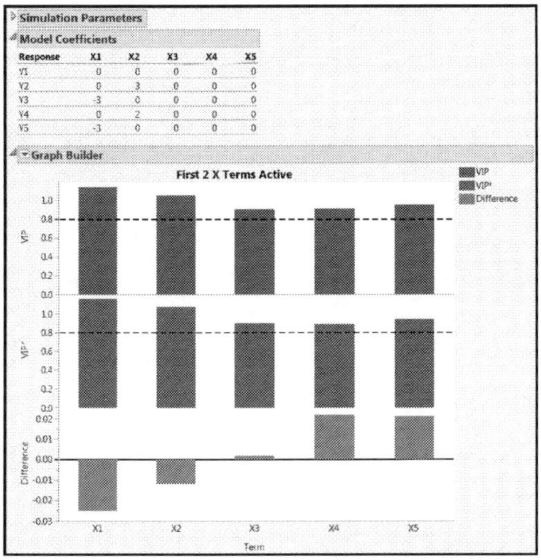

Response	X1	X2	X3	X4	X5
Y1	0	0	0	0	0
Y2	0	3	0	0	0
Y3	-3	0	0	0	0
Y4	0	2	0	0	0
Y5	-3	0	0	0	0

Run the script under several conditions to see the effect, and then close any open reports before continuing.

The Standardize X Option

This option is available only on the Fit Model launch window, when Partial Least Squares is selected as the personality. It is of interest if you construct model terms from the columns in your data table. Suppose that you have two columns, X_1 and X_2, and that you are interested in including interaction or polynomial terms. For an example, suppose that you add the term $X_1 * X_2$ as an effect in the Fit Model launch window.

The **Center** and **Scale** options construct the product using the raw measurements in the columns X_1 and X_2. If only these options are selected, the product is centered and scaled, so that the variable that enters the PLS calculation is

$$\frac{X_1 X_2 - mean(X_1 X_2)}{stdev(X_1 X_2)}$$

But, if you center and scale your columns, you might want to form polynomial terms from centered and scaled columns, rather than from the original data values. When you enter the term $X_1 * X_2$ in the Fit Model launch window, the **Standardize X** option inserts this term into the model:

$$\left(\frac{X_1 - mean(X_1)}{stdev(X_1)}\right) * \left(\frac{X_2 - mean(X_2)}{stdev(X_2)}\right)$$

This product is then centered and scaled based on selection of the **Center** and **Scale** options.

The three options **Center**, **Scale**, and **Standardize X** are checked by default in the Fit Model launch window. If all of your effects are main effects, the models fit with and without the **Standardize X** option are identical.

Determining the Number of Factors

Cross Validation: How JMP Does It

Cross validation is based on the *Predicted Residual Sums of Squares* (PRESS) statistic. This section illustrates the calculation of this statistic.

Suppose that you specify **KFold** as the **Validation Method** and set the **Number of Folds** to h. We will describe how the **Root Mean PRESS** values, found in the **KFold Cross Validation** report, are calculated. Note that, when you run the model with **KFold** as the **Validation Method**, under the report for the suggested fit, you are given the option to **Save Columns > Save Validation**. This saves a column containing an identifier for the holdout set to which a given row belongs.

Specify a number of factors, say a factors. The **Root Mean PRESS** value for a factors can be calculated as follows:

1. Exclude the observations in the ith holdout set.
2. Fit a model with a factors to the remaining observations, specifying **None** as the **Validation Method**.
3. Save the prediction formulas for this model by selecting **Save Columns > Save Prediction Formula**.
4. For each of the k **Y**s, calculate PRESS values for that response for the observations in the ith fold as follows: Compute the squared difference between the observed value and the predicted value (the squared prediction error), and divide the result by the variance for the entire response column.
5. Sum the means of these values across the h holdout sets and divide the sum of these means by the number of folds minus one. Call the result PRESS(Y).
6. The **Root Mean PRESS** is the square root of the mean of the PRESS(Y) values across the k responses.

The data table WaterQuality_PRESSCalc.jmp illustrates the calculation of the **Root Mean PRESS** for a NIPALS model with 2 factors. You can open this table by clicking on the correct link in the master journal. For simplicity, we have selected two of the responses from the WaterQuality.jmp data table in Chapter 7, HAB and RICH, and 47 rows from the original data table.

Run the PLS Fit script. This script contains a random seed, so that you can obtain the same results as are shown in the data table. The **KFold Cross Validation with K=2 and Method = NIPALS** report shows the **Root Mean PRESS** values that appear in Figure 2.

Figure A1.2: Cross Validation Report

KFold Cross Validation with K= 2 and Method=NIPALS

Number of factors	Root Mean PRESS	van der Voet T²	Prob > van der Voet T²
0	1.441992	0.000000	1.0000
1	1.507991	1.763501	0.4610
2	1.512265	1.537996	0.5200
3	1.692740	2.993765	0.2470
4	1.636567	3.845911	0.1650
5	1.667821	4.031805	0.1350
6	1.711594	4.148925	0.1230
7	1.715969	4.380944	0.1020

The data table WaterQuality_PRESSCalc.jmp contains steps for the calculation of the **Root Mean PRESS** value for **Number of factors** equal to **2**, namely 1.512265. Note that the column called Validation in the data table is precisely the validation column associated with this specific report. To verify this, from the **NIPALS Fit with 2 Factors** red triangle menu, select **Save Columns > Save Validation**. Once you have verified that you obtain the same fold assignments, you can delete the column you have added (Validation 2).

Run the script Predictions Fold 1. This script excludes the observations in Validation fold 2, and fits a two-factor NIPALS model on only the data in Validation fold 1. It saves the prediction formulas in columns called Pred Formula HAB_2 and Pred Formula RICH_2, where the "2" indicates that these are applied to the test data in fold 2. Run Predictions Fold 2. This script saves prediction formulas built using the data in fold 2 in columns called Pred Formula HAB_1 and Pred Formula RICH_1.

Now run the script PRESS Calculations. This saves formulas to the data table that accomplish the calculations described in steps 4 through 6 above. The final column, RM PRESS 2 Factors, shows the value 1.512265, which is the value shown in Figure A1.2.

For details about the van der Voet test, see the *SAS/STAT 9.3 User's Guide* and search for "van der Voet".

Appendix: Simulation Studies

Introduction 249
The Bias-Variance Tradeoff in PLS 250
    Introduction 250
    Two Simple Examples 250
    Motivation 254
    The Simulation Study 255
    Results and Discussion 257
    Conclusion 261
A Utility Script to Compare PLS1 and PLS2 261
Using PLS for Variable Selection 263
    Introduction 263
    Structure of the Study 264
    The Simulation 267
    Computation of Result Measures 268
    Results 270
    Conclusion 280

Introduction

Simulation studies can be very useful when trying to understand the efficacy and nuances of different approaches to statistical modeling. This is because, in order to generate the data to analyze in the first place, the true model must be known. This appendix presents two simulation studies that provide insight on how PLS is likely to work in practice. Bear in mind, though, that any true model used to generate data is at best only representative of reality. So simulation studies always have a restricted, but hopefully relevant, scope.

The first study addresses the bias-variance tradeoff in PLS. This study is of particular interest when your modeling objective is prediction. The second study builds on the material in More on VIPs in Appendix 1. It is of relevance when you are using PLS in an explanatory context, and specifically when your goal is variable selection. The code for the simulations is written entirely in JSL.

The Bias-Variance Tradeoff in PLS

Introduction

The prediction error of a statistical model can be expressed in terms of three components:

1. The **noise**, which is irreducible and intrinsic to any statistical approach.

2. The **squared bias**, where bias is the difference between the average of the predictions and the true value.

3. The **variance**, which is the mean squared difference between the predictions and their average.

Generally, whichever modeling approach is used, increasing model complexity (measured in an appropriate way) reduces the squared bias but increases the variance. In the context of supervised learning, this phenomenon is called the *bias-variance tradeoff*. Finding an appropriate balance between the contributions from the squared bias and the variance can lead to smaller prediction errors. When the modeling goal is prediction, it is important to find this balance.

Details of the bias-variance tradeoff in the predictive modeling context can be found in Hastie et al. 2001. But the essentials are easily grasped through the following examples.

Two Simple Examples

We have already seen a simple example of the bias-variance tradeoff in the case of polynomial regression in our discussion in "Underfitting and Overfitting: A Simulation" in Chapter 2. In that section's example, the order of the polynomial chosen for the fit is the measure of model complexity.

Run the script **PolyRegr2.jsl** found in the Appendix 2 material in the master journal file. This script shows the underlying true model with a blue curve, and shows the fitted polynomial model in red. By running this script several times to produce different random realizations of the true model, you can investigate fits for a simple model (say **Order** = 1) and a complex model (say **Order** = 7).

Figures A2.1 and A2.2 show representative plots. For **Order** = 1 you see that the red curve, which is a line in this case, changes very little from run to run (has low variance). But for **Order** = 7, the curve changes greatly (has high variance) because it follows the points as they move around. Open the **Model Generalization to New Data** report to see how the RMSE for the test set increases with the order of the polynomial.

Figure A2.1: Simple Model: Low Variance, High Bias Fits

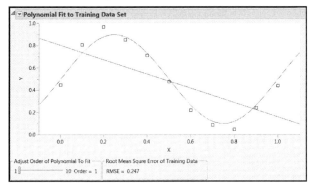

Figure A2.2: Complex Model: High Variance, Low Bias Fits

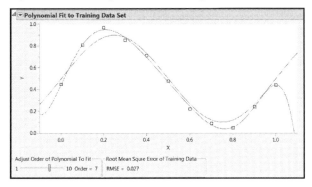

A second example is provided by the script BiasAndVarianceOf2DClassifiers.jsl. Run this script by clicking on its link in the master journal. The simulation involves two continuous predictors, x1 and x2, and a single categorical response, Output Class. Output Class assumes only two values, "0" (colored green) and "1" (colored red).

We can view the responses as located on a plane, with points colored appropriately. The underlying, true model is the generative Gaussian process, which produces some intermingling of red and green points. Running this script produces a random realization from the generative Gaussian.

The script uses two approaches to predict class membership for the simulated data:

- A nominal logistic regression of Output Class in terms of x1 and x2 (Fit Model with the Nominal Logistic personality).

- A kth nearest neighbor approach, in which a fine grid is defined, and the color of each grid point is defined by voting over a neighborhood defined by the value of k.

The script plots the data, along with the classification into red and green that is provided by the two different model fits. Use the slider in the **High Variance, Low Bias Decision Boundary** report to set the number of nearest neighbors used in defining the classification. Figures A2.3 and A2.4 show some typical output.

Figure A2.3: Decision Boundary from Nominal Logistic Regression

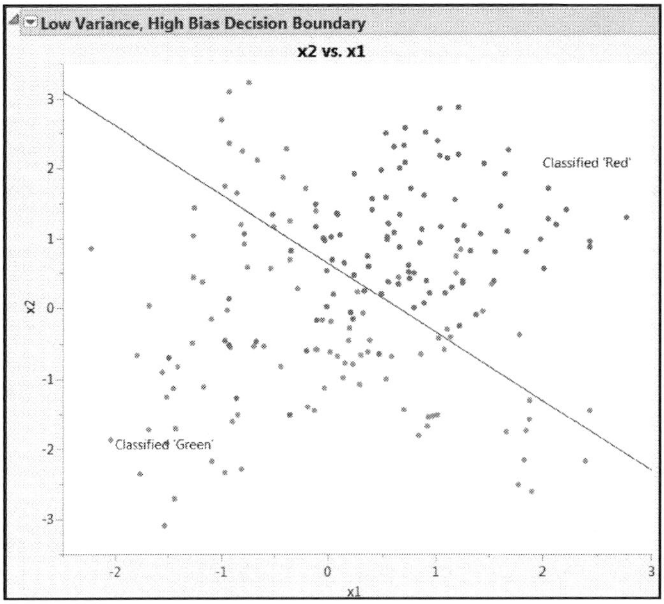

Figure A2.4: Decision Boundary with 5 Nearest Neighbors

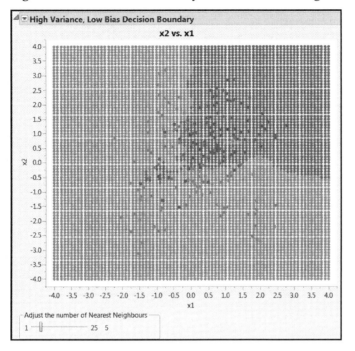

Because of its nature, the decision boundary for the nominal logistic fit is a line. Points on one side of the line are classified as being red; those on the other are classified as green. Figure A2.3 shows that many points are misclassified. This model may not be complex enough to provide low prediction error.

The nearest neighbor approach, on the other hand, can provide a *wiggly* decision boundary that appears better able to separate the red and green points. (See Figure A2.4, where misclassified points are marked with an **x**.). However, note that small values of k can provide extremely wiggly boundaries and high variance, low bias, fits. As k increases, causing more neighbors to enter the average used to predict class membership, the complexity of the model decreases, resulting in decreased variance and increased bias. The goal is to find the balance between variance and bias that minimizes prediction error.

Running BiasAndVarianceOf2DClassifiers.jsl several times gives a better appreciation of what data can come from the same underlying model, and how the two different fitting approaches predict class membership.

Motivation

The PLS literature sometimes alludes to a PLS analysis involving a single **Y** with the acronym *PLS1* and a PLS analysis involving multiple **Y**s with the acronym *PLS2*. For example, the water quality example in Chapter 7 is an example of a PLS2 analysis, since it requires the prediction of four responses (**Y**s).

Recall that the **Y**s are laboratory measurements of water quality from 86 samples taken at specific locations in the Savannah River basin, whereas the 26 **X**s are features of the terrain local to the sample locations, extracted from remote sensing data. Figure 7.36, reproduced here as Figure A2.5 for convenience, shows actual versus predicted values for the test set of six rows. (Recall that two test set rows were missing values on HAB.)

Figure A2.5: Predictions on Test Set for PLS2 Pruned Model

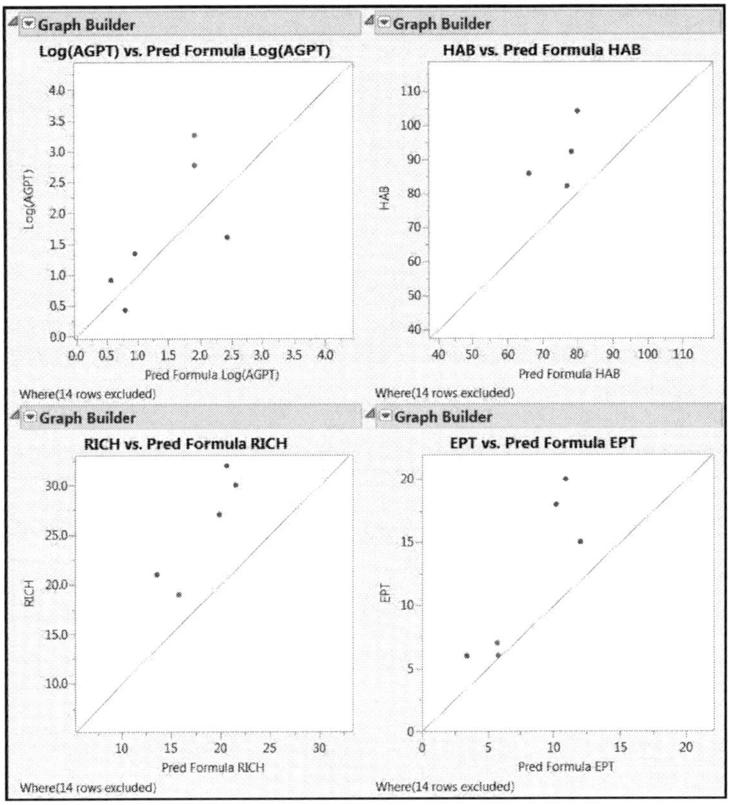

The apparent bias in Figure A2.5, coupled with the general importance of the bias-variance tradeoff as described in this chapter's "Introduction", motivates the simulation study presented next. This study also presents an opportunity to study differences between the PLS approaches (PLS1 or PLS2) and the fitting algorithms used (NIPALS and SIMPLS). However, because the results given by NIPALS and SIMPLS are identical in the case of a single **Y** (PLS1), there are actually only three combinations to investigate, not four.

The Simulation Study

The Model

Our underlying model consists of five **Y**s and five **X**s. Twenty sets of values for the five **X**s were simulated with a particular correlation structure to mimic data that is typically analyzed using PLS. The correlation structure of the **X**s is described in detail in the second simulation study in this appendix.

Twenty sets of true values for the **Y**s were generated from known linear functions of the **X**s defined by the coefficients shown in Figure A2.6. This process produces a data set with 20 rows and 10 columns. Note that each of the five **X**s is *active* in the sense that each contributes to some degree to the prediction of each of the **Y**s. These coefficients are saved in ModelCoefficients.jmp, which we link to in the master journal for your reference.

Figure A2.6: Coefficients for the True PLS Model

	Response	Coeff of X1	Coeff of X2	Coeff of X3	Coeff of X4	Coeff of X5
1	Y1	-1.310	1.011	1.369	2.274	1.409
2	Y2	-2.013	-0.639	0.329	0.647	-1.032
3	Y3	0.490	-0.839	1.844	-1.244	0.810
4	Y4	1.158	-0.881	-0.576	-1.622	1.260
5	Y5	-0.214	-0.401	-0.545	0.669	-0.104

The Simulated Xs

The 20 simulated values for X1 through X5 and the corresponding true **Y**s are given in PLSSimulation_OneModel.jmp, also included in the master journal. The correlation structure induced between the **Y**s as a result of these functional dependencies is shown in Figure A2.7. The figure also shows the correlations among the **X**s and between the **X**s and the **Y**s. Run the script **All Correlations** to obtain these reports.

Figure A2.7: Correlations in the True PLS Model with Twenty Observations

Multivariate

Correlations

	X1	X2	X3	X4	X5	True Y1	True Y2	True Y3	True Y4	True Y5
X1	1.0000	0.6856	0.4253	0.2855	0.0086	0.1247	-0.9221	0.3571	-0.0772	-0.7873
X2	0.6856	1.0000	0.8196	0.6994	0.3668	0.6578	-0.8386	0.4530	-0.6874	-0.8343
X3	0.4253	0.8196	1.0000	0.8523	0.5996	0.8720	-0.6698	0.7703	-0.8104	-0.6791
X4	0.2855	0.6994	0.8523	1.0000	0.7824	0.9597	-0.5470	0.5026	-0.8056	-0.3159
X5	0.0086	0.3668	0.5996	0.7824	1.0000	0.8567	-0.3605	0.5879	-0.3994	-0.0758
True Y1	0.1247	0.6578	0.8720	0.9597	0.8567	1.0000	-0.4544	0.5943	-0.8082	-0.3095
True Y2	-0.9221	-0.8386	-0.6698	-0.5470	-0.3605	-0.4544	1.0000	-0.5845	0.2788	0.8445
True Y3	0.3571	0.4530	0.7703	0.5026	0.5879	0.5943	-0.5845	1.0000	-0.2978	-0.6064
True Y4	-0.0772	-0.6874	-0.8104	-0.8056	-0.3994	-0.8082	0.2788	-0.2978	1.0000	0.3472
True Y5	-0.7873	-0.8343	-0.6791	-0.3159	-0.0758	-0.3095	0.8445	-0.6064	0.3472	1.0000

Color Map On Correlations

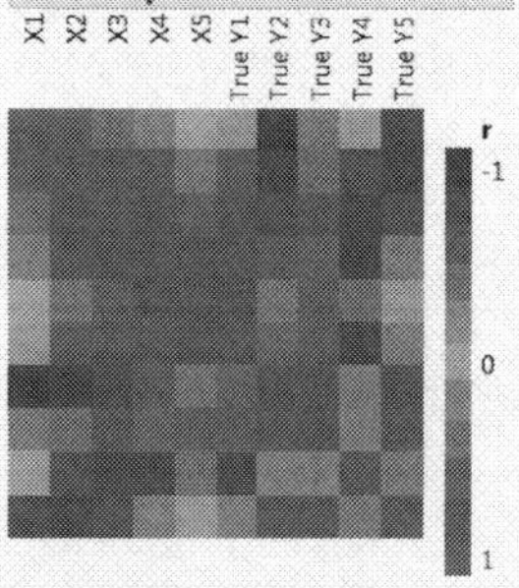

The Study

Recall that PLS1 refers to fitting each response using a PLS model, while PLS2 refers to fitting a single PLS model to all responses. In our case, PLS1 indicates that five PLS models were fit, one for each response.

PLS fits, using one to five factors, were obtained using each of the following three approaches in turn:

1. NIPALS with PLS1 (identical to PLS1 with SIMPLS)
2. NIPALS with PLS2
3. SIMPLS with PLS2

The results of the simulations are given in Results_OneModel.jmp, available in the master journal. Four factors define the study:

- NObs (number of observations): 4 or 20
- Standard deviation Sigma: 1.00, 1.50, and 2.00
- Method and Variant at three levels: NIPALS with PLS1, NIPALS with PLS2, and SIMPLS with PLS2

- Number of Factors to be fit: 1–5

For each value of Sigma, 3,000 random realizations of the true model were generated for the case where NObs = 20. The **X**s with the fixed values shown in the table PLSSimulation_OneModel.jmp were used. Random normal noise with mean zero and standard deviation Sigma was added to each **Y**. Each simulation produced five **Y**s with random values for analysis. This process provided three sets of 1,000 simulations, one for each of NIPALS with PLS1, NIPALS with PLS2, and SIMPLS with PLS2.

For each simulation previously obtained, *four* of the 20 rows of generated data were randomly selected to create data for the NObs = 4 scenarios. This added another three sets of 1,000 simulations. These scenarios mimic *wide* data, as the resulting data had 10 columns and only four rows.

PLS analyses were then run on the six sets of simulations according to the specified values of Method and Variant and Number of Factors, resulting in a total of 90 (6 x 3 x 5) combinations. The summary measures for all of these simulations are given in rows 1 through 90 of Results_OneModel.jmp.

Results and Discussion

For each combination of Method and Variant, Number of Factors, Sigma, and NObs conditions, the Squared Bias, Variance, and mean squared prediction error (MSPE) for the 1,000 simulations were calculated. These values were averaged over the five responses, giving one value of each of these measures for each set of conditions, as shown in the table Results_OneModel.jmp.

For the case where NObs = 20, Figures A2.8, A2.9, and A2.10 compare the three Method and Variant approaches in terms of their MSPE, Squared Bias, and Variance, respectively, across Number of Factors. To explore these measures for the case where only four rows are used, run the scripts MSPE Across Factors, Squared Bias Across Factors, and Variance Across Factors, setting NObs in the **Local Data Filter** to 4.

Figure A2.8: Mean Squared Prediction Error, NObs = 20

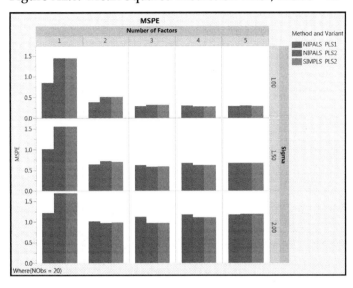

Figure A2.9: Squared Bias, NObs = 20

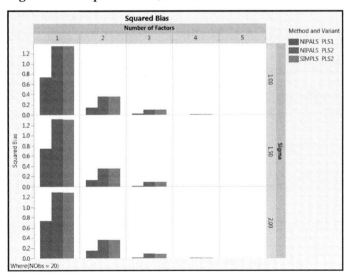

Figure A2.10: Variance, NObs = 20

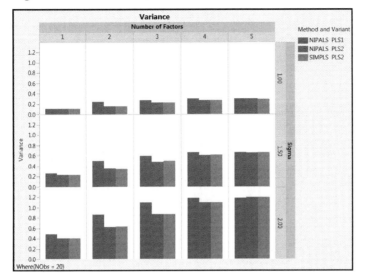

You see the following:

1. MSPE increases with the value of sigma, as expected. The increase is due to increased variance. The squared bias is unaffected by sigma.

2. As the number of factors increases, squared bias decreases and variance increases. MSPE remains fairly constant for two or more factors.

3. NIPALS PLS2 and SIMPLS PLS2 have similar values for all three of MSPE, squared bias, and variance, regardless of the number of factors extracted (compare the red and green bars).

4. For the one and two factor models, MSPE for NIPALS PLS1 is generally lower than for both PLS2 methods. NIPALS PLS1 has lower squared bias than PLS2, and slightly higher variance.

When NObs = 4, these remarks apply as well, in general terms.

Figure A2.8 shows that the MSPE values for NIPALS and SIMPLS with PLS2 are essentially identical when NObs = 20. Figure A2.11 shows how closely the MSPE values track, for both NObs = 4 and NObs = 20. Sigma is colored using a blue to red intensity scale, with blue representing Sigma = 1 and red representing Sigma = 3. Note the expected higher prediction error for the fits where NObs = 4 and for larger values of Sigma. (The figure can be constructed by running the Graph Builder script in the table

Results_OneModel_PLS2Split.jmp, found in the master journal.)

Figure A2.11: Comparison of Mean Squared Prediction Error for PLS2

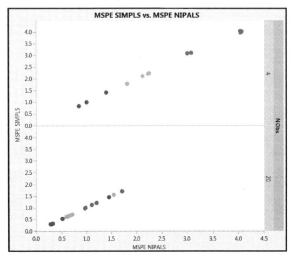

For prediction, the model with the lowest MSPE is preferable. Figure A2.12 compares the three approaches for various sample sizes and values of sigma. For this underlying model at least, the fitted PLS1 model with the lowest MSPE for a given NObs, Method, and Sigma never has more latent variables than the corresponding PLS2 model. Note that our current simulation does not investigate the use of different cross validation schemes to identify the optimal number of latent variables. This would be an interesting extension.

Figure A2.12: Best Models

	NObs	Sigma	Method	Variant	MSPE	Number of Factors
1	4	1.00	NIPALS	PLS1	0.8661	1
2	4	1.00	NIPALS	PLS2	0.8377	2
3	4	1.00	SIMPLS	PLS2	0.8370	2
4	4	1.50	NIPALS	PLS1	1.7472	1
5	4	1.50	NIPALS	PLS2	1.8139	2
6	4	1.50	SIMPLS	PLS2	1.7820	2
7	4	2.00	NIPALS	PLS1	3.0184	1
8	4	2.00	NIPALS	PLS2	3.0048	1
9	4	2.00	SIMPLS	PLS2	3.0844	1
10	20	1.00	NIPALS	PLS1	0.2900	3
11	20	1.00	NIPALS	PLS2	0.2839	4
12	20	1.00	SIMPLS	PLS2	0.2802	4
13	20	1.50	NIPALS	PLS1	0.6172	3
14	20	1.50	NIPALS	PLS2	0.5878	3
15	20	1.50	SIMPLS	PLS2	0.5996	3
16	20	2.00	NIPALS	PLS1	1.0146	2
17	20	2.00	NIPALS	PLS2	0.9747	2
18	20	2.00	SIMPLS	PLS2	0.9752	2

Conclusion

Even though it is widely used in many diverse application areas, PLS still has unresolved aspects. Results from the simulation study described here suggest that, in the PLS2 context, models fit with NIPALS and SIMPLS have very similar predictive ability. Depending on the number of factors and the degree of noise, relative to the coefficients of the true model, PLS1 might result in lower or higher mean squared prediction errors than PLS2, but the differences are small. An important finding is that PLS1 seems to produce less bias than PLS2.

A Utility Script to Compare PLS1 and PLS2

The previous simulation study shows that when you have multiple **Y**s in your study there can be differences between the results of the PLS1 and PLS2 approaches. This is not surprising given the way that PLS works.

If you have multiple **Y**s, conducting PLS1 can be burdensome because you have to manually analyze each **Y** on its own. If you use PLS2, once you have finished your modeling, it can be interesting to see how your results stack up against the PLS1 equivalents.

The utility script ComparePLS1andPLS2.jsl is designed to help you decide whether to use PLS1 or PLS2 in analyzing your data. You can run this script by clicking on its link in the master journal.

Before opening the script, you need to make the data table of interest the current data table. Figure A2.13 shows how to populate the launch window for the data in the table WaterQuality_BlueRidge.jmp found in the section "A First PLS Model for the Blue Ridge Ecoregion" in Chapter 7. Setting **Display Wrap** to 2 in the launch window gives two plots per row in the final report. Click the **Help** button for other details.

Figure A2.13: Populated Launch Window for Comparing PLS1 and PLS2 Predictions

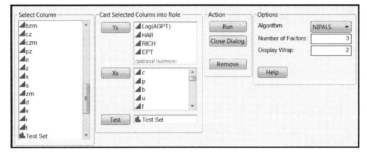

With the window populated as shown in Figure A2.13, click **Run** to obtain the report shown in Figure A2.14. Using this report, you can easily compare the predictions made for each response by PLS1 and by PLS2. In each pair of plots, the top plot shows PLS1 predicted values, while the bottom plot shows PLS2 predicted values. PLS2 seems to have more bias for HAB than does PLS1. For the other responses, both methods seem to behave similarly: the plots for HAB, RICH, and EPT suggest that both PLS1 and PLS2 result in some bias.

Figure A2.14: Comparison of PLS1 and PLS2 Predictions Using the Utility Script

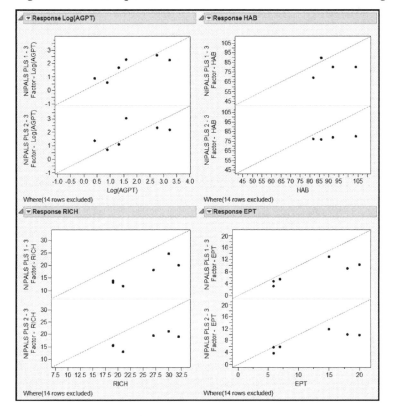

Using PLS for Variable Selection

Introduction

Selecting only the active variables from a set of potential predictors can be of immense explanatory value. In the case of MLR, one standard approach is to use stepwise regression. Even within the stepwise regression framework, though, there is considerable complexity and sophistication, leading to a number of alternative prescriptions.

In the case of PLS, the VIP values and standardized coefficients of model terms are often used to reduce the number of terms in the model. If the goal is prediction, such a reduction is reasonable if it does not compromise predictive ability. But if the goal is explanation, care is required. PLS, by its nature, is of value when correlations exist among the Xs. The script Multicollinearity.jsl, discussed in the section "The Effect of Correlation Among Predictors: A Simulation" in Chapter 2, shows the difficulties that can arise when trying to interpret coefficients under such conditions in the MLR setting. The data table PLSvsTrueModel.jmp in the section "An Example Exploring Prediction" in Chapter 4 illustrates a consequence of variable selection in PLS when two predictors are highly correlated.

In Appendix 1, the script Compare_NIPALS_VIP_and_VIPStar.jsl in the section "More on VIPs" compared two alternative approaches to computing VIP values in a NIPALS fit. The simulation study here extends this considerably by exploring the efficacy of different prescriptions for variable selection via PLS.

As mentioned previously, simulations can only be indicative rather than exhaustive. In terms of representing real studies, here are several important things to consider:

1. The numbers of **X**s, **Y**s, and observations.
2. The correlation structure of the **X**s.
3. The number of active **X**s, and how the active and inactive **X**s are arranged within the correlation structure.
4. The functional form that links the active **X**s to the **Y**s.
5. The level of intrinsic noise.

The simulation study we describe next attempts to deal with all these aspects in a limited but useful way. All simulations were done using JSL directly, not using any JMP platforms.

Structure of the Study

The Numbers of Xs, Ys, and Observations

The study considered 6 **Y**s and 20 **X**s, with 20 observations for each combination.

The Correlation Structure of the Xs

Two types of correlation structure were simulated, **cType** = 1 and **cType** = 2.

The correlation structure **cType** = 1 is the same as the correlation structure used in the first simulation study. For some visual intuition, run the script cType_1_CorrelationStructure.jsl by clicking on the corresponding link in the master journal (Figure A2.15). The parameter **rhoX** controls the rate at which correlation values fall off as you move away from the leading diagonal. The parameter **nX** is the number of **X**s, and **divisorFac** is another parameter that is set to 0.25 in the simulation. The correlations themselves are given in the report following the plots. For this correlation structure, the correlation coefficients are always positive.

Figure A2.15: X Correlation Structure Type 1

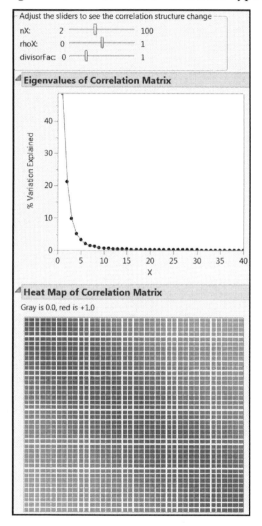

You can explore the second type of correlation structure, **cType** = 2, using the script cType_2_CorrelationStructure.jsl. Click on the corresponding link in the master journal. Again, **nX** is the number of **X**s. One or two peaks can be selected from the drop-down list. The parameter **peakSize** controls the proportion of the **X**s affected by the peak.

Figure A2.16 shows a situation with two peaks. The correlation values fall off as you move away from the leading diagonal and the secondary peak. In this structure, negative correlations can occur.

Figure A2.16: X Correlation Structure Type 2

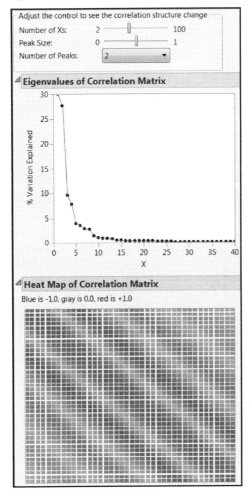

Once the type of correlation structure was selected, along with the associated parameter values, the correlation matrix for the 20 predictors (**nX** = 20) was computed. Then, we generated 20 observations from a mean zero multivariate Gaussian distribution with that correlation matrix. This gives the required **X** matrix.

Active Xs and Their Arrangement

The total number of **X**s was 20 (**nX** = 20). The number of active **X**s, **nXactive,** was assigned values of 2, 4, 6, 8, or 10. In each simulation a "coefficient budget", **coeffTot**, varying between 1 and 10 in integer increments, was used to constrain the **nXactive** terms in their relationship with the **Y**s. The locations of the **nXactive** terms in the list of **nX** terms were controlled by another variable, **coeffType**. The variable **coeffType** assumed one of four possible values: Equal Left, Unequal Left, Equal Random, and Unequal Random. As might be expected, "Left" implies that all the **nXactive** terms were placed first in the list of 20. "Random" indicates that all the **nXactive** terms were placed at random in the list of 20. "Equal" indicates that the available budget was allocated equally among the active terms. Finally, "Unequal" indicates that the budget was allocated unequally, specifically according to a binomial-like distribution, among the active terms. This is a generalization of the scheme used in Chong and Jun (2005).

The Functional Relationship

Once coefficients were defined for the **nXactive** terms by particular choices for **coeffTot** and **coeffType**, true values for all the **Y**s were calculated as linear functions of the **X**s using these coefficients. This gave a **YTrue** matrix to be paired with the **X** matrix generated earlier. Call the combined matrix **TrueData**.

The Simulation

Once **True Data** had been constructed, random normal noise with standard deviation **Sigma** was added to each **Y**. **Sigma** values of 0.01, 0.1, 1.0, 2.0, 3.0 were used.

For each row of **TrueData** and for each value of **Sigma**, 100 realizations were generated from **TrueData**. We call the resulting matrix, which we used for PLS modeling, **Data**. Each realization was then fit using the NIPALS and SIMPLS algorithms, and **a** = 1, 2, 3, 4, 5, 6 factors were extracted. **Cutoff** values of 0.8 and 1.0 were applied to obtain results.

The simulation runs conducted made up a full-factorial design of all the relevant variables. Focusing on the case of **cType** = 1, the simulation factors and levels were as shown in Table A2.1.

Table A2.1: Full Factorial Design for Correlation Type 1

Simulation Factor	Levels	Number of Levels
rhoX	0.0 to 0.9 in increments of 0.1	10
nXactive	2 to 10 in increments of 2	5
coeffTot	1 to 10 in increments of 1	10
coeffType	Equal Random, Unequal Random, Equal Left, Unequal Left	4
Sigma	0.01, 0.1, 1.0, 2.0, 3.0	5
a	1 to 6 in increments of 1	6
Cutoff	0.8, 1.0	2

For the case **cType** = 1, this gives 120,000 different treatment combinations to consider. For the case **cType** = 2, the factors and levels were the same as in Table 1, except that **rhoX** was replaced by **peakSize** (0.05 to 0.50 in 0.05 increments, giving 10 levels) and **nPeak** (1 or 2, giving 2 levels). So in this case there were 240,000 different treatment combinations.

Thus the design matrix for the simulation contains 360,000 rows. For each of the 30,000 combinations consisting of a row of **TrueData** and a value of **Sigma**, 100 simulations were generated. For each of these, the number of factors, **a**, was varied, and the two **Cutoff** values were applied.

Computation of Result Measures

For each row of **TrueData**, we know by construction which specific factors are active, and also the true coefficient values involved. For each simulated data set, we attempt to identify these active factors from the fits using **Cutoff** values of 0.8 and 1.0 for the NIPALS VIP, NIPALS VIP*, and SIMPLS VIP values. We also compare the predicted values and estimated coefficients. Specifically, we calculate the following:

- The true positive and false positive rates for NIPALS VIP, SIMPLS VIP, and VIP* for both cut-off values. The true positive rate for a simulation is the number of active factors that exceed the cut-off divided by the number of active factors. The false positive rate is the number of inactive factors that exceed the cut-off divided by the number of inactive factors.

- The overall error rates for NIPALS VIP, SIMPLS VIP, and VIP* for both cut-off values. The overall error rate for a simulation is the number of active factors that do not exceed the cut-off plus the number of inactive factors that do exceed the cut-off, divided by the total number of factors (20).
- The maximum absolute difference of the NIPALS and SIMPLS predictions.

More specifically, we calculate values for the following metrics:

1. **TP NIPALS VIP** (true positive rate for terms identified using NIPALS VIP values).
2. **FP NIPALS VIP** (false positive rate for terms identified using NIPALS VIP values).
3. Overall **Error Rate VIP**.
4. **TP NIPALS VIP*** (true positive rate for terms identified using NIPALS VIP* values).
5. **FP NIPALS VIP*** (false positive rate for terms identified using NIPALS VIP* values).
6. Overall **Error Rate VIP***.
7. **TP SIMPLS VIP** (true positive rate for terms identified using SIMPLS VIP values).
8. **FP SIMPLS VIP** (false positive rate for terms identified using SIMPLS VIP values).
9. Overall **Error Rate SIMPLS VIP**.
10. **MAD Beta BetaSIMP** (maximum absolute difference between NIPALS coefficients and SIMPLS coefficients).
11. **MAD PredY PredYSIMP** (maximum absolute difference between NIPALS predicted values and SIMPLS predicted values).
12. **MAD VIP VIPStar** (maximum absolute difference between NIPALS VIP and NIPALS VIP* values).
13. **MAD VIP VIPSIMP** (maximum absolute difference between NIPALS VIP and SIMPLS VIP values).

14. **MAD VIPStar VIPSIMP** (maximum absolute difference between NIPALS VIP* and SIMPLS VIP values).

For each of the 360,000 treatment combinations, these metrics were averaged over the 100 simulations.

Results

Comparison of VIP Values

We begin by exploring the differences between the NIPALS and SIMPLS VIP values. This is interesting in relation to the commonly used threshold values of 0.8 and 1.0 for detecting active terms.

Figures A2.17 and A2.18 show mean absolute difference results for **cType** = 1 and **cType** = 2, respectively, using a common vertical scale. Note that, when **a** = 1, the VIP values are identical because no deflation is involved. For **a** > 1, as **Sigma** increases, the differences increase. The impact of the different correlation types of the **X**s is small. The blue lines link mean values.

Figure A2.17: Differences between NIPALS VIP and SIMPLS VIP for Type 1 Correlation

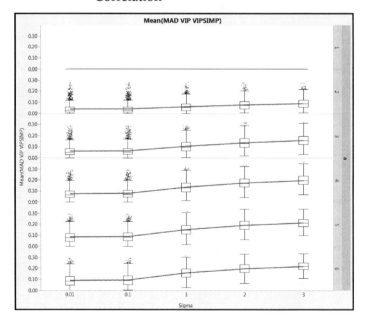

Figure A2.18: Differences between NIPALS VIP and SIMPLS VIP for Type 2 Correlation

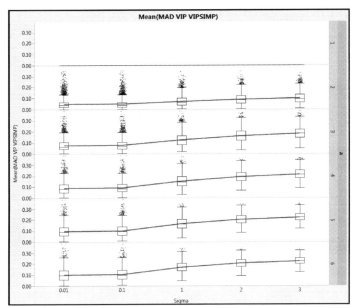

Figures A2.19 and A2.20 show mean absolute difference results between the NIPALS VIP* and the SIMPLS VIP values (using the same scales as Figures A2.17 and A2.18. Based on the details in the section "More on VIPs" in Appendix 1, we might expect these differences to be smaller, and these plots verify this. The **X** correlation type has even less impact than for the NIPALS and SIMPLS VIP comparison.

Figure A2.19: Differences between NIPALS VIP* and SIMPLS VIP for Type 1 Correlation

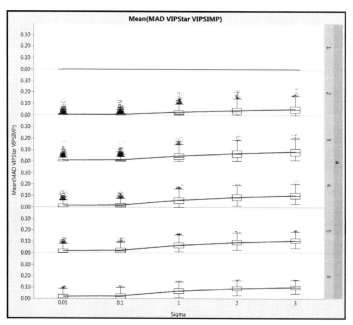

Figure A2.20: Differences between NIPALS VIP* and SIMPLS VIP for Type 2 Correlation

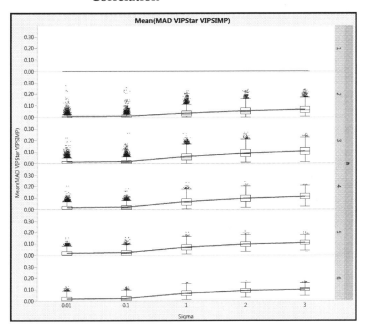

Comparison of Variable Selection Error Rates

For the variable selection problem, **Error Rate VIP**, **Error Rate VIP***, and **Error Rate SIMPLS** are perhaps the most important metrics. Each is the sum of the number of false positives and false negatives identified by the corresponding method (**NIPALS VIP**, **NIPALS VIP***, and **SIMPLS VIP**), divided by the number of Xs (20). Without more information about the specific study objectives, it is difficult to say whether false positives or false negatives are more damaging, so we confine ourselves to discussing the results for these overall error rates.

Furthermore, knowing that NIPALS VIP* values are fairly close to SIMPLS VIP values, and because the JMP PLS platform does not calculate VIP* values directly, we further confine the rest of the discussion to comparing the NIPALS VIP error rate to the SIMPLS VIP error rate (**Error Rate VIP** and **Error Rate SIMPLS**).

A tabular comparison of mean error rates based on NIPALS and SIMPLS VIPs and for **cType** = 1 and **cType** = 2 is shown in Figure A2.21. The overall error rates for the 1.0 cut-off are considerably lower than for the 0.8 cut-off. Overall error rates computed using SIMPLS VIPs tend to be lower than those using NIPALS VIPs. This particular study suggests that SIMPLS VIPs outperform NIPALS VIPs for both types of errors, but especially for false positives. We caution, though, that this is a limited study. Also, one needs to better understand the false positive/false negative tradeoff before making a determination relative to which cut-off or method is best.

Also note that, for the cut-off value of 0.8, the error rates for **cType** = 2 are always less than or equal to those for **cType** = 1. When the cut-off is 1.0, the error rates are fairly similar.

Figure A2.21: Comparison of Error Rates for NIPALS and SIMPLS VIPs

cType = 1

| | | CutOff | | | |
| | | 0.8 | | 1 | |
Sigma	a	NIPALS VIP Mean	SIMPLS VIP Mean	NIPALS VIP Mean	SIMPLS VIP Mean
0.01	1	0.40	0.40	0.27	0.27
	2	0.43	0.41	0.23	0.22
	3	0.44	0.42	0.22	0.21
	4	0.45	0.42	0.21	0.21
	5	0.46	0.42	0.21	0.21
	6	0.47	0.43	0.21	0.21
0.1	1	0.40	0.40	0.27	0.27
	2	0.44	0.41	0.23	0.22
	3	0.45	0.42	0.22	0.21
	4	0.46	0.42	0.22	0.21
	5	0.47	0.43	0.22	0.21
	6	0.47	0.43	0.22	0.21
1	1	0.43	0.43	0.32	0.32
	2	0.49	0.46	0.29	0.29
	3	0.51	0.48	0.30	0.29
	4	0.53	0.49	0.30	0.29
	5	0.55	0.50	0.31	0.30
	6	0.56	0.51	0.31	0.30
2	1	0.47	0.47	0.36	0.36
	2	0.52	0.50	0.35	0.34
	3	0.55	0.52	0.36	0.35
	4	0.57	0.53	0.37	0.36
	5	0.59	0.54	0.38	0.36
	6	0.60	0.55	0.38	0.37
3	1	0.49	0.49	0.39	0.39
	2	0.54	0.52	0.39	0.38
	3	0.56	0.54	0.40	0.38
	4	0.58	0.55	0.41	0.39
	5	0.60	0.56	0.41	0.40
	6	0.61	0.57	0.42	0.40

cType = 2

| | | CutOff | | | |
| | | 0.8 | | 1 | |
Sigma	a	NIPALS VIP Mean	SIMPLS VIP Mean	NIPALS VIP Mean	SIMPLS VIP Mean
0.01	1	0.36	0.36	0.27	0.27
	2	0.39	0.37	0.24	0.23
	3	0.40	0.38	0.23	0.22
	4	0.42	0.38	0.22	0.21
	5	0.43	0.39	0.22	0.21
	6	0.44	0.39	0.22	0.21
0.1	1	0.36	0.36	0.27	0.27
	2	0.39	0.37	0.25	0.24
	3	0.41	0.38	0.23	0.22
	4	0.42	0.38	0.23	0.22
	5	0.43	0.39	0.23	0.22
	6	0.44	0.40	0.23	0.22
1	1	0.39	0.39	0.32	0.32
	2	0.45	0.43	0.31	0.29
	3	0.48	0.45	0.30	0.29
	4	0.50	0.46	0.30	0.29
	5	0.52	0.48	0.31	0.30
	6	0.54	0.49	0.31	0.30
2	1	0.43	0.43	0.36	0.36
	2	0.49	0.47	0.36	0.35
	3	0.53	0.49	0.36	0.35
	4	0.55	0.51	0.37	0.35
	5	0.57	0.53	0.37	0.36
	6	0.59	0.54	0.38	0.36
3	1	0.45	0.45	0.38	0.38
	2	0.52	0.50	0.39	0.38
	3	0.55	0.52	0.40	0.38
	4	0.58	0.54	0.40	0.39
	5	0.59	0.55	0.41	0.39
	6	0.61	0.56	0.41	0.40

Next, we illustrate how error rates vary with **Sigma**. Figures A2.22 and A2.23 show **Error Rate VIP** and **Error Rate SIMPLS** for **cType** = 1, and Figures A2.24 and A2.25 show the same metrics for **cType** = 2. Generally, the distributions are wide, but the advantage of using a cut-off value of 1.0 rather than 0.8 relative to overall error rate is clear. As **Sigma** increases, the mean error rate increases (shown by the blue line) and its variability decreases (shown by both the interquartile range and the total range).

Figure A2.22: Error Rate for NIPALS VIP for Type 1 Correlation

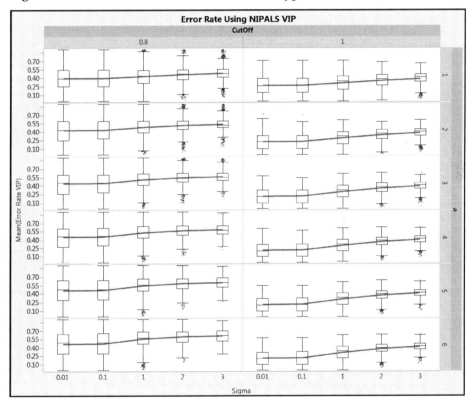

Figure A2.23: Error Rate for SIMPLS VIP for Type 1 Correlation

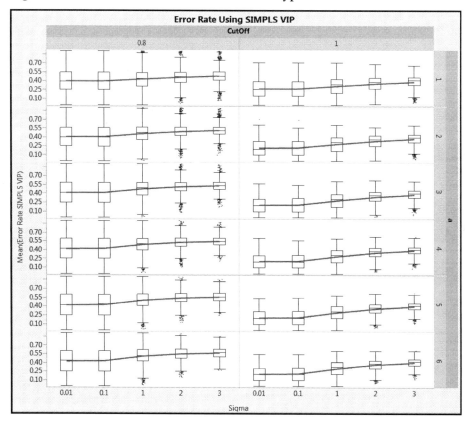

Figure A2.24: Error Rate for NIPALS VIP for Type 2 Correlation

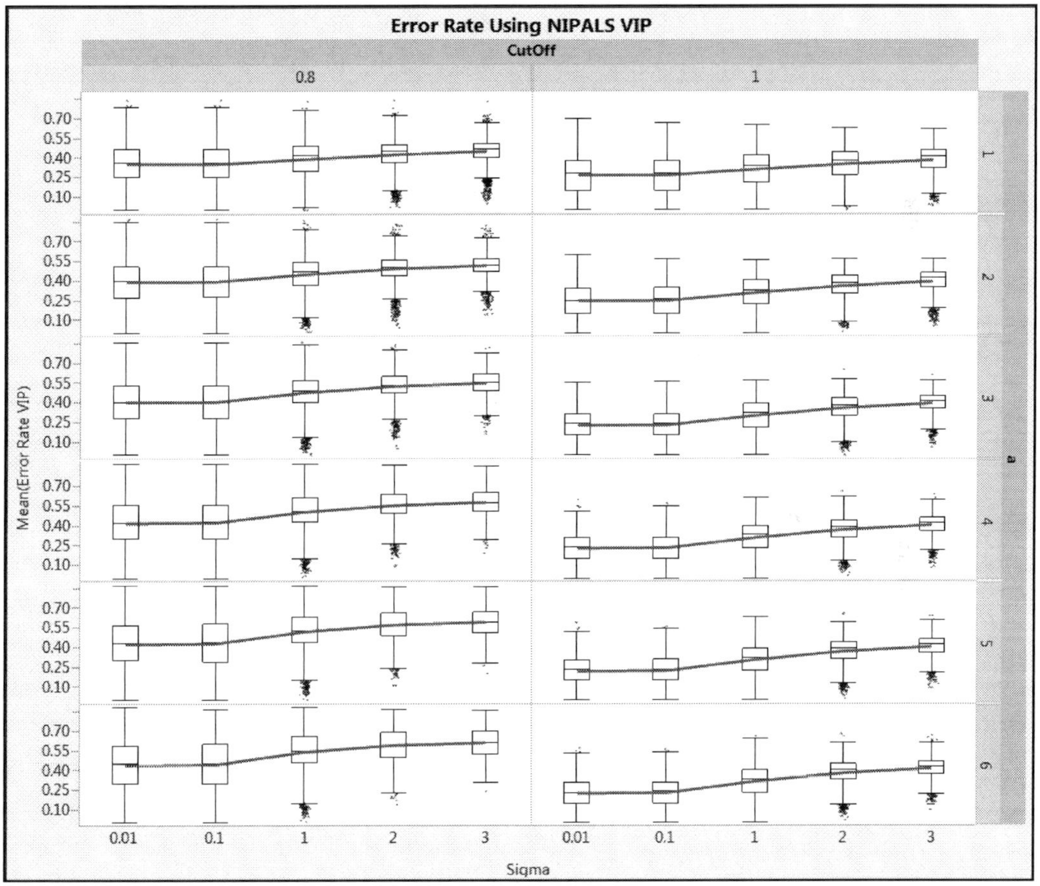

Figure A2.25: Error Rate for SIMPLS VIP for Type 2 Correlation

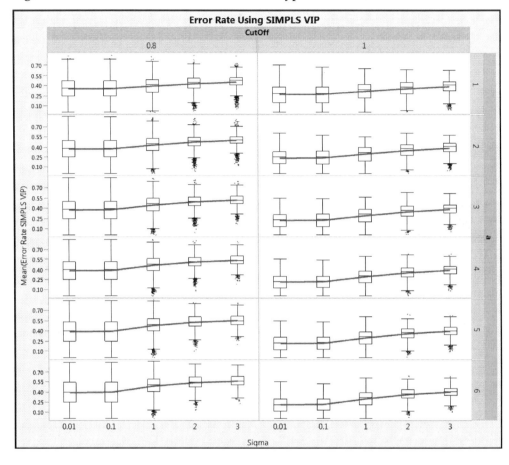

Let's delve a bit more deeply into if and how the error rates change with other important factors such as **nXactive**, **coeffTot**, and **coeffType**.

Figure A2.26 shows NIPALS VIP error rates for **cType** = 1, but using histograms rather than box plots as we did in Figure A2.22. In addition, all rows corresponding to **coeffTtype == Equal Left** and **Unequal Left** have been selected, so that the conditional distribution of the "Left" arrangement of coefficients is highlighted. This also allows a comparison with the two "Random" arrangements of coefficients.

Figure A2.26 clearly shows that, for small values of **Sigma**, the "Left" arrangement produces lower error rates than the "Random" arrangement. These differences lessen as **Sigma** increases. Figure A2.27 shows NIPALS VIP error rates for **cType** = 2. This figure leads to the same conclusions as does Figure A2.26.

Figure A2.26: Effect of CoeffType on NIPALS VIP Error Rate for Type 1 Correlation

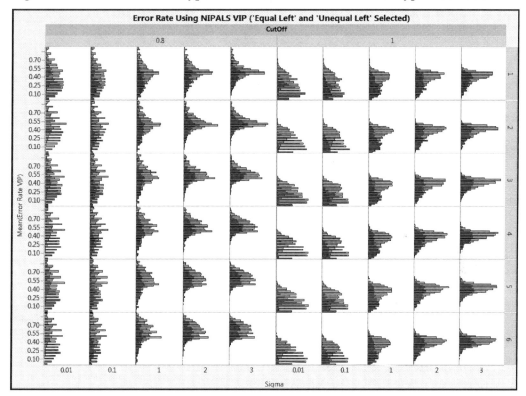

Figure A2.27: Effect of CoeffType on NIPALS VIP Error Rate for Type 2 Correlation

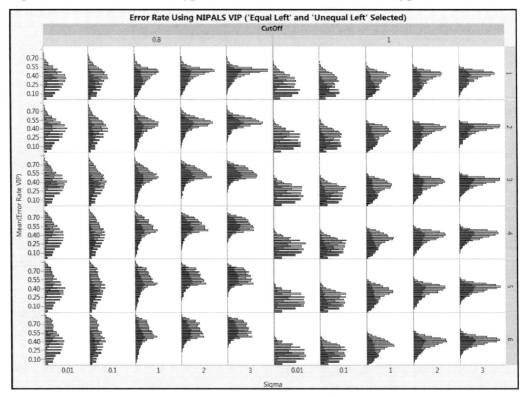

Conclusion

In spite of the complexity revealed by the simulation, the major findings are clear. The SIMPLS VIP and NIPALS VIP* appear to outperform the NIPALS VIP by a slight margin. The cut-off value has a major impact on the error rate, with a cut-off of 1.0 resulting in lower overall error rates than a cut-off of 0.8. We recommend using the 1.0 cut-off if you are interested in the overall error rate. If you are interested in maximizing true positives, you should use the 0.8 cut-off. Not surprisingly, the error rates are also affected by the correlation structure of the Xs and how the active factors are dispositioned within the correlation structure.

Variable selection in PLS is a complex problem. We suggest caution and encourage you to compare your results with those obtained by alternative approaches when these exist.

A secondary conclusion from our simulation study, for which we did not present details, is that NIPALS and SIMPLS tend to give similar predictions.

References

Andersson, M. (2009), "A Comparison of Nine PLS1 Algorithms," *Journal of Chemometrics*, 23:10, 518–529.

Arlot, S., and Celisse, A. (2010), "A Survey of Cross-Validation Procedures for Model Selection," *Statistics Surveys*, 4, pp. 40–79.

Belsley, D.A. (1991), *Conditioning Diagnostics: Collinearity and Weak Data in Regression*, New York: Wiley-Interscience.

Chakrapani, C. (2004), *Statistics in Market Research*, London: Arnold.

Chatfield, C. (1995), *Problem Solving: A Statistician's Guide*, 2d ed., New York: Chapman & Hall/CRC.

Chong, I.G., and Jun, C.H. (2005), "Performance of Some Variable Selection Methods When Multicollinearity is Present," *Chemometrics and Intelligent Laboratory Systems,* 78:1-2, 103–112.

De Jong, S. (1993), "SIMPLS: An Alternative Approach to Partial Least Squares Regression," *Chemometrics and Intelligent Laboratory Systems,* 18:3, 251–263.

Dijkstra, T.K. (2010), "Latent Variables and Indices: Herman Wold's Basic Design and Partial Least Squares." In Vinzi, V. E., Chin, W.W., Henseler, J., and Wang, H. (eds.), *Handbook of Partial Least Squares: Concepts, Methods and Applications*, Berlin: Springer, Chapter 1, 23–46.

Draper, N.R., and Smith, H. (1998), *Applied Regression Analysis*, 3d ed., New York: John Wiley & Sons, Inc.

Eriksson, L., Johansson, E., Kettaneh-Wold, N., Trygg, J., Wikstrom, C., and Wold, S. (2006), *Multi- and Megavariate Data Analysis Basic Principles and Applications (Part I)*, Chapter 4, Umetrics.

Hastie, T., Tibshirani, R., and Friedman, J. (2001), *The Elements of Statistical Learning: Data Mining, Inference, and Prediction,* New York: Springer.

Hellberg, S., Sjöström, M., and Wold, S. (1986), "The Prediction of Bradykinin Potentiating Potency of Pentapeptides. An Example of a Peptide Quantitative Structure-activity Relationship," *Acta Chemica Scandinavica* B, 40, 135–140.

Hoskuldsson, A. (1988), "PLS Regression Methods," *Journal of Chemometrics,* 2:3, 211–228.

Kalivas, J.H. (1997), "Two Data Sets of Near Infrared Spectra," *Chemometrics and Intelligent Laboratory Systems,* 37:2, 255–259.

Kourti, T., and MacGregor, J.F. (1996), "Multivariate SPC Methods for Process and Product Monitoring," *Journal of Quality Technology,* 28:4, 409–428.

Mateos-Aparicio, G. (2011), "Partial Least Squares (PLS) Methods: Origins, Evolution, and Application to Social Sciences," *Communications in Statistics – Theory and Methods,* 40:13, 2305–2317.

Microsoft Research (2009), The Fourth Paradigm: Data-Intensive Scientific Discovery, Hey, T., Tansley, S., and Tolle, K. (eds.), Microsoft Corporation.

Nash, M.S., and Chaloud, D.J. (2011), "Partial Least Square Analyses of Landscape and Surface Water Biota Associations in the Savannah River Basin," *International Scholarly Research Network, ISRN Ecology,* Vol. 2011, Article ID 571749, 11 pages.

Nomikos, P., and MacGregor, J.F. (1995), "Multivariate SPC Charts for Monitoring Batch Processes," *Technometrics,* 37:1, 41–59.

O'Mahony, M. (1986), *Sensory Evaluation of Food: Statistical Methods and Procedures,* New York: Marcel Dekker.

Pérez-Enciso, M., and Tenenhaus, M. (2003), "Prediction of Clinical Outcome with Microarray Data: A Partial Least Squares Discriminant Analysis (PLS-DA) Approach," *Human Genetics,* 112:5-6, 581–592.

SAS Institute Inc. (2011), *SAS/STAT 9.3 User's Guide,* "The PLS Procedure," Cary, NC: SAS Institute Inc.

SAS Institute Inc. (2011), *Using JMP,* Cary, NC: SAS Institute Inc. For the most current version, go to http://www.jmp.com/support/downloads/documentation.shtml.

Sjöström, M., and Wold, S. (1985), "A Multivariate Study of the Relationship Between the Genetic Code and the Physical-Chemical Properties of Amino Acids," *Journal of Molecular Evolution,* 22:3, 272–277.

Tabachnick, B.G., Fidell, L.S. (2001), *Using Multivariate Statistics,* Allyn and Bacon, Boston.

Tracy, N.D., Young, J.C., and Mason, R.L. (1992), "Multivariate Control Charts for Individual Observations," *Journal of Quality Technology,* 24:2, 88–95.

Tobias, R.D. (1995), "An Introduction to Partial Least Squares Regression," *Proceedings of the Twentieth Annual SAS Users Group International Conference,* Cary, NC: SAS Institute Inc.

Ufkes, J.G.R., Visser, B.J., Heuver, G., and Van der Meer, C. (1978), "Structure-Activity Relationships of Bradykinin Potentiating Peptides," *European Journal of Pharmacology,* 50:2, 119–122.

Ufkes, J.G.R., Visser, B.J., Heuver, G., Wynne, H.J., and Van der Meer, C. (1982), "Further Studies on the Structure-Activity Relationships of Bradykinin-Potentiating Peptides," *European Journal of Pharmacology,* 79:1-2, 155–158.

van den Wollenberg, A.L. (1977), "Redundancy Analysis - An Alternative for Canonical Correlation Analysis," *Psychometrika,* 42:2, 207–219.

van der Voet, H. (1994), "Comparing the Predictive Accuracy of Models Using a Simple Randomization Test," *Chemometrics and Intelligent Laboratory Systems,* 25:2, 313–323.

Wold, H. (1966), "Estimation of principal components and related models by iterative least squares." In Krishnaiah, P.R. (ed.), *Multivariate Analysis; Proceedings.* New York: Academic Press, 391–420.

Wold, H. (1980). "Soft Modelling: Intermediate between Traditional Model Building and Data Analysis," *Mathematical Statistics* (Banach Center Publications, Warsaw), 6, 333–346.

Wold, S. (1995), "PLS for Multivariate Linear Modeling," in *Chemometric Methods in Molecular Design. Methods and Principles in Medicinal Chemistry,* van de Waterbeemd, H. (ed.), New York: VCH. [SAS Library Note: The second part of the title (Methods and Principles in Medicinal Chemistry) is actually the name of the Series.]

Wold, S., Johansson, E., and Cocchi, M. (1993), "PLS – Partial Least-Squares Projections to Latent Structures," in *3D QSAR in Drug Design: Theory Methods and Applications,* Vol.1, Kubinyi, H. (ed.), 523–550, Dordrecht, The Netherlands: Kluwer.

Wold, S., Sjostrom, M., Eriksson, L. (2001), "PLS-Regression: A Basic Tool of Chemometrics," *Chemometrics and Intelligent Laboratory Systems,* 58:2, 109–130.

Index

A

Actual by Predicted Plot report 89, 95, 162, 168–169, 171–172, 197, 213
actual values, compared with predicted values 96–99, 170–172, 179–180
AGPT, transforming 148–150
analysis, performing 79–81, 156–158, 175–176
Andersson, M. 222
applied statistics 1
Arlot, S. 67
assessing models using test set 133–136

B

Baking Bread That People Like example
  about 183–184
  combined model 215–219
  data 184–186
  first stage model 187–202
  second stage model 202–215
Belsley, D.A. 22
Bias-Variance Tradeoff in PLS
  about 250, 261
  examples 250–253
  motivation 254–255
  results and discussion 257–261
  simulation study 255–257
BigClassCVDemo.jmp 67–68
bivariate distributions 205, 207
Blue Ridge Ecoregion, PLS models for 173–180
Bradykinin potentiating activity 76, 103
Bread.jmp 185–186, 203, 215–219

C

Cars example 11–15
CarsSmall.jmp 11–15
Cauchy-Schwarz Inequality 233
Celisse, A. 67

centering
  data 3–4
  example of 28–31
  in PLS 37–38
Chackrapani, C. 184
Chaloud, D.J. 140, 142, 155, 167, 179, 181, 182
Chatfield, C. 2
Chong, I.G. 267
Coefficient Plots 131–132, 181, 212
choosing number of factors 65–71
column vector 12–15
combined model, for Baking Bread That People Like example 215–219
Compare_NIPALS_VIP_and_VIPStar.jsl script 264
ComparePLS1andPLS2.jsl script 262–263
comparing
  actual values to predicted values 96–99, 170–172, 179–180
  residuals 99–101
  stage one models for Baking Bread That People Like example 200–202
  variable selection error rates 273–280
  VIP values 270–273
confidence ellipses,
  Scatterplot Matrix 188
  X Score Scatterplot Matrix 87
correlation
  effect of among predictors 18–23
  for Ys and Xs in PLS 38
  with factors, loadings as measures of 235–236, 243
  structure of Xs 264–266
Correlations report 189
covariance 32, 232–233
creating
  formula 148
  plots of individual spectra 111–112
  stacked data tables 109–111
  subsets 173–174
  test set indicator columns 107–108

cross validation 18, 65–66, 66–67, 246–248
    See also k-fold cross validation
    See also leave-one-out cross validation

D

data
    Baking Bread That People Like example 184–186
    centering 3–4
    contextual nature of 2
    diversity of 2
    imputing missing 146–147
    initial visualization 77–79
    performance on 96–104
    Predicting Biological Activity example 76–79
    Predicting Octane Rating of Gasoline example 106–108
    reviewing 174–175
    scaling 3–4
    transforming 3–4
    viewing 108–116
    Water Quality in Savannah River Basin example 140–141
data filter 46–48, 57, 113–116, 124–125, 257
de Jong, S. 72, 237
deflation 225
design matrix 14
diagnostics
    Predicting Biological Activity example 95–96
    Predicting Octane Rating of Gasoline example 121–125
    pruned PLS model for Savannah River basin 168–169
Diagnostics Plots 88–89, 122, 160–162, 193–194, 210–211
differences, by ecoregion 150–155
Dijkstra, T.K. 71
dimensionality reduction 34–36
DimensionalityReduction.jsl script 31–34, 34–36

dimensions, of matrices 15
Distance Plots 96, 123
distances, to X and Y models 242
distributions, Water Quality in Savannah River Basin example 147–148
Draper, N.R. 15

E

eigenvalues 26–27, 222–224, 233
eigenvectors 27, 223–224
Eriksson, L. 4, 28, 89, 96, 163
examples
    See also specific examples
    Bias-Variance Tradeoff in PLS 250–253
    Cars 11–15
    centering 28–31
    of PLS analysis 4–10
    prediction 59–64
    scaling 28–31
    scores and loadings 54–55
excluding test set 116–117
expected value 19
exploratory data analysis, in multivariate studies 31–34
extracting factors 50–51

F

factor loadings 52, 224
factors
    choosing number of 65–71
    determining number of, using cross validation 246–248
    extracting 50–51
    3-D scatterplots for one 46
"fat matrices" 106
Fidell, L.S. 48
fingerprint 42
first principal component 26, 36
Fit Model Launch Window 5–6
Fit Model Platform 15

Fitting PLS models 117–118, 120, 136–137, 190–195, 207–208
formula editor window 217
Friedman, J. 23, 39, 54, 250

G

Gasoline.jmp 106–116
generative Gaussian process 252

H

Hastie, T. 23, 39, 54, 250
Hellberg, S. 76, 77
Heuver, G. 76, 77, 79, 103
holdout cross validation
    method 66
    set 247
Hoskuldsson, A. 232

I

identity matrix 16
imputing missing data 146–147
initial data visualization
    Baking Bread That People Like example 77–79
    Water Quality in Savannah River Basin example 144–145
initial reports 118–120
inner relation regression 52
inputs 1
inverse of a square matrix 16

J

JMP customizations for SIMPLS 241
JMP Pro
    Fit Model launch 5–6
    KFold validation 7

    Validation Methods 69–70
Johansson, E. 4, 28, 89, 96, 163
Jun, C.H. 267

K

Kalivas, J.H. 106
Kettaneh-Wold, N. 4, 28, 89, 96, 163
k-fold cross validation 66, 67–68
KFold Cross Validation report 119, 167–168, 208, 209
Kourti, T. 126

L

leave-one-out cross validation 66–67, 190–192, 208–209
Leave-One-Out report 190–192, 208–209
left singular vectors 222
linked subset 174
loading matrix 52, 83–85
loading plots 27, 85–86, 127–128
loadings
    as measures of correlation with factors 235–236, 243
    PLS 50–59
    properties of 232
LoWarp.jmp 28–31
lurking variables 103

M

MacGregor, J.F. 126
Make Model Using VIP 92, 130, 136, 195
Make Model Using Selection 92, 93, 130, 164, 166, 178, 212
Mason, R.L. 126
Mateos-Aparicio, G. 71
matrices
    dimensions of 15
    "fat matrices" 106

identity 16
loading 52, 83–85
scatterplot 83–87, 125–127
singular value decomposition of 222–223
matrix algebra 222
maximization of covariance 232–233
mean squared prediction error (MSPE) 257–261
Microsoft Research 2
missing response values, Water Quality in Savannah River Basin example 145–146
Missing Value Imputation report 147, 159–160
MLR
    See multiple linear regression (MLR)
Model Coefficients report 193, 210
Model Comparison Summary report 119, 158–159, 191–192, 208, 209
model fitting
    for Baking Bread That People Like example 195–197, 197–199, 212–215
    PLS model for Blue Ridge Ecoregion 178–180
    pruned PLS model for Savannah River Basin example 166–168
modeling 1–2
models
    assessing using test set 133–136
    fitting 117–118, 120, 136–137, 207–208
    in terms of X scores 52, 228–229, 241
    testing 9–10
    for X and Y 52–53, 228–229, 241
MSPE (mean squared prediction error) 257–261
multicollinearity 18–23
Multicollinearity.jsl script 18–23, 263–264
multiple linear regression (MLR)
    Cars example 11–15
    effect of correlation among predictors 18–23
    estimating coefficients 15–16
    overfitting 16–18
    underfitting 16–18
multivariate studies, exploratory data analysis in 31–34

multivariate technique, PLS as a 38–39

N

Nash, M.S. 140, 142, 155, 167, 179, 181, 182
NIPALS algorithm
    about 71–72, 226–228
    computational results 228–231
    extracting factors 50
    models in terms of X scores 52
    models in terms of Xs 53
    notation 225–226
    one-factor model 60–63
    properties of 231–237
    two-factor model 63–64
NIPALS Fit report 159–160, 176–178
NIPALS Fit with 1 Factors report 158–163, 191–192, 193, 196
NIPALS Fit with 2 Factors report 208, 209, 247–248
NIPALS Fit with 3 Factors report 7–8, 120
noise 14
Nomikos, P. 126
nonlinear iterative partial least squares algorithm
    See NIPALS algorithm
notation
    for NIPALS algorithm 225–226
    for SIMPLS algorithm 238–240
number of factors 246–248

O

O'Mahony, M. 184
opening formula editor window 217
optimization criterion, SIMPLS algorithm 237
outputs 1
overfitting 16–18

P

parameters 15
partial least squares (PLS)
    See also variable selection
    about 1–2
    algorithms 224–225
    analysis example 4–10
    as a multivariate technique 38–39
    centering in 37–38
    compared with PCA 49–50
    how it works 45–49
    loadings 50–59
    models 155–181, 173–180
    models for Blue Ridge Ecoregion 173–180
    models for predicting biological activity 79–96
    models for predicting octane ratings of gasoline 116–138
    report 44, 81–82, 158–159, 191–192, 208–212
    reasons for using 39–45
    scaling in 37–38
    scores 50–59
    overview 72–73
    in today's world 2–3
    variable reduction in 89–90
Partial Least Squares Model Launch window 7
Partial Least Squares report 158–159, 191–192, 208–212
PCA
    See principal components analysis (PCA)
PCA platform 27
PCR (Principal Components Regression) 39, 223–224
Penta.jmp 76–77
Percent Variation Explained for X Effects 230, 242
Percent Variation Explained for Y Responses 230, 242
Percent Variation Explained report 121, 137, 192, 209
Pérez-Enciso, M. 89

performing analysis 79–81, 96–104. 156–158, 175–176
plots
    construction for individual spectra 111–112
    diagnostics 88–89, 122
    loading 27, 85–86, 127–128
    variable importance 90–93
PLS
    See partial least squares (PLS)
PLS platform 69–71
PLS procedure 77
PLS Report 81–82
PLS1 models 222
PLS2 models 222
PLSGeometry.jsl script 45–49
PLS_PCA.jsl script 49–50
PLSScoresAndLoadings.jmp 54–55
PLSvsTrueModel.jmp 59–60
PolyRegr.jsl script 16–18
PolyRegr2.jsl script 250–253
Predicted Residual Sums of Squares (PRESS) statistic 246–248
predicted values, compared with actual values 96–99, 170–172, 179–180
Predicting Biological Activity example
    about 75–76
    data 76–79
    first PLS model 79–93
    performance on data from second study 96–104
    pruned PLS model 93–96
Predicting Octane Rating of Gasoline example
    about 106
    data 106–108
    first PLS model 116–120
    pruned model 136–138
    second PLS model 120–136
    viewing data 108–116
prediction
    example using simulation 59–64
    formulas, saving 8–9, 60–64, 169–170
Prediction Profiler 201–202, 214–215, 218–219

predictors, effect of correlation among 18–23, 59–64
PRESS (Predicted Residual Sums of Squares) statistic 246–248
principal components 224
principal components analysis (PCA)
    about 25–27, 223–224
    compared with PLS 49–50
    dimensionality reduction via 34–36
Principal Components Regression (PCR) 39, 223–224
Profiler
    comparing via the 201–202
    viewing 213–215, 218–219
projection method 48
projection to latent structures 48
properties
    of loadings 232
    of NIPALS algorithm 231–237
    of SIMPLS algorithm 237–238
    of scores 232
    shared by NIPALS and SIMPLS 53–54

R

regression
    inner relation in PLS 52
    stepwise 263
regression coefficients 15, 130, 234–235
regression parameters 12
regularization techniques 23
reports
    Actual by Predicted Plot 89, 95, 162, 168–169, 171–172, 197, 213
    Coefficient Plots 131–132, 181, 212
    Diagnostics Plots 88–89, 122, 160–162, 193–194, 210–211
    Distance Plots 96, 123
    initial 118–120
    KFold Cross Validation 119, 167–168, 208, 209
    Leave-One-Out 190–192, 208–209
    Loading Plots 83–86, 127–128
    Missing Value Imputation 147, 159–160
    Model Coefficients 193, 210
    Model Comparison Summary 119, 158–159, 191–192, 208, 209
    NIPALS Fit with 1 Factors 158–163, 191–192, 193, 196
    NIPALS Fit with 2 Factors 208, 209, 247–248
    NIPALS Fit with 3 Factors 7–8, 120
    Partial Least Squares (PLS) 44, 81–82, 158–159, 191–192, 208–212
    Percent Variation Explained 121, 137, 192, 209
    Profiler 201–202, 213–215, 218–219
    Residual by Predicted Plot 89, 95, 122, 168, 197
    Score Scatterplot Matrices 86–87, 125–127
    SIMPLS Fit with 2 Factors 82–83
    Stepwise Regression Control 198–199
    T Square Plot 123, 160–161
    Variable Importance Plot 44–45, 90–91, 129, 131–132, 165
    VIP vs Coefficients Plots 91–93, 130–132, 136, 163–166, 177–178, 194–195, 212
    X-Y Scores Plots 82–83, 120–121, 159, 192, 196, 209
Residual by Predicted Plot report 89, 95, 122, 168, 197
residuals
    about 14–15, 34
    comparing 99–101
right singular vectors 223
RMSE (Root Mean Square Error) 17–18, 67–68
Root Mean PRESS (Predicted Residual Sum of Squares) statistic 69, 119–120, 167, 192, 209, 246–249
Root Mean Square Error (RMSE) 17–18, 67–68
Rose, David 184

S

SAS/STAT 9.3 User's Guide 77, 248
saving prediction formulas 8–9, 96, 133, 169–170, 248

scaling
    data 3–4
    example of 28–31
    in PLS 37–38
scatterplot matrices
    loading matrix 83–86
    scoring 86–87, 125–127
score vectors 224
scores
    PLS (partial least squares) 50–59
    properties of 232
Score Scatterplot Matrices report 86–87, 125–127
second principal component 26
Sensory Evaluation of Food: Statistical Methods and Procedures (O'Mahony) 184
SIMPLS algorithm
    about 71–72, 237, 240–246
    extracting factors 50
    fits 64
    implications for 237–238
    models in terms of Xs 53
    notation 238–240
    optimization criterion 237
SIMPLS Fit report 82–83
simulation studies
    about 249–250
    Bias-Variance Tradeoff in PLS 250–261
    overfitting 16–18
    underfitting 16–18
    using PLS for variable selection 263–280
    Utility Script to Compare PLS2 and PLS2 261–263
singular value decomposition of a matrix 222–223
singular values 222–223
Sjöström, M. 76, 77
Smith, H. 15
Solubility.jmp 25–27, 34
Spearheads.jmp 4–5, 66
spectra
    combined 113–116
    constructing plots of individual 111–112
    individual 112–113
spectral decomposition, relationship to singular value decomposition 223
SpectralData.jsl script 40–45
SS(YModel) 242
stacked data tables, creating 109–111
stage one MLR model, for Baking Bread That People Like example 197–200
stage one pruned model, for Baking Bread That People Like example 195–197
stage two MLR model, for Baking Bread That People Like example 212–215
stage two PLS model, for Baking Bread That People Like example 207–208
Standardize X option 246
statistical models 1–2
Statistically Inspired Modification of the PLS Method
    See SIMPLS algorithm
Statistics in Market Research (Chackrapani) 184
stepwise regression 189, 263
Stepwise Regression Control report 198–199
Stratified sample, creating 155–156
subsets, creating 173–174
sum of squares
    for contribution of factor f to X model 230
    for factor f to Y model 229
    for X 242
    for Y 242

T

T Square Plot 123, 160–161
Tabachnick, B.G. 48
Tenenhaus, M. 89
test set
    about 5
    assessing models using 133–136
    creating indicator columns 107–108
    creating stratified sample 155–156
    excluding 116–117
testing models 9–10
3-D scatterplots, for one factor 46

Tibshirani, R. 23, 39, 54, 250
Tobias, R.D. 38
Tracy, N.D. 126
training set 5, 65
transforming
    creating a column formula 148–149
    through a launch window 148–150
    weights 236–237
transforming data 3–4
transpose 16
Trygg, J. 4, 28, 89, 96, 163

U

Ufkes, J.G.R. 76, 77, 79, 103
underfitting 16–18
univariate distributions 204
Utility Script to Compare PLS2 and PLS2 261–263

V

validation
    k-fold cross validation 66, 67–68, 119, 167–168, 208, 209
    leave-one-out cross validation 66–67, 190–192, 208–209
    in PLS platform 69–71, 246–249
validation set 65
van den Wollenberg, A.L. 39
van der Meer, C. 76, 77, 79, 103
van der Voet, H. 69, 119
van der Voet tests 69–70, 137, 167–168, 248
Variable Importance for the Projection (VIP) statistic
    See VIP (Variable Importance for the Projection) statistic
Variable Importance Plot report 44–45, 90–91, 129, 131–132, 165
variable selection
    about 64, 189, 263–264, 280

comparing error rates in simulation study 273–280
computation of result measures for simulation study 268–270
results of simulation study 270–280
simulation 267–268
structure of simulation study 264–267
variables
    comparing selection error rates 273–280
    lurking 103
    reduction in PLS 89–90
    relationships between 187–188
    visualizing two at a time 152–154
variance, bias toward X directions with high variance 234
viewing
    data 108–116
    Profiler 201–202, 213–215, 218–219
    VIPs for spectral data 131
VIP (Variable Importance for the Projection) statistic
    about 129–133
    comparing values 270–273
    for *i*th predictor 230–231, 243–244
    variable reduction in PLS 89
    viewing for spectral data 131
VIP vs Coefficients Plots report 91–93, 130–132, 136, 163–166, 177–178, 194–195, 212
VIP* 231, 244–245, 268–273, 280
Visser, B.J. 76, 77, 79, 103
visualizing
    data 77–79
    two variables at a time 152–154
    Ys and Xs 202–207

W

Water Quality in Savannah River Basin example
    about 140–141
    data 141–155
    defined 140
    first PLS model 155–166

pruned PLS model 166–172
WaterQuality2.jmp 155–156
WaterQuality2_Train.jmp 156–158
WaterQuality_BlueRidge.jmp 174–175
WaterQuality.jmp 140–141
WaterQuality_PRESSCalc.jmp 247–248
weights, transforming 236–237
Wikstrom, C. 28
Wold, H. 71
Wold, S. 4, 28, 71, 76, 77, 89, 96, 129, 163, 231
Wynne, H.J. 76, 77, 103

X

X
- Active, in simulation 267
- correlation structure of, in simulation 264–266
- models for 52–53, 241
- models in terms of scores 50–52
- properties of weights 232
- sums of squares for 242

X-Y Scores Plots report 82–83, 120–121, 159, 192, 196, 209

Y

Young, J.C. 126
Y
- models for 52–53, 241
- sums of squares for 242

Symbols

* matrix multiplication 12–13
β column vector of regression parameters 12–13, 15–16
ε column vector of errors 12–13, 15–16
Σ correlation matrix 38–39

Made in the USA
Middletown, DE
14 January 2017